AF782520

THEORY AND DECISION LIBRARY

General Editors: W. Leinfellner (*Vienna*) and G. Eberlein (*Munich*)

Series A: Philosophy and Methodology of the Social Sciences

Series B: Mathematical and Statistical Methods

Series C: Game Theory, Mathematical Programming and Operations Research

Series D: System Theory, Knowledge Engineering and Problem Solving

SERIES C: GAME THEORY, MATHEMATICAL PROGRAMMING AND OPERATIONS RESEARCH

VOLUME 39

Editor-in-Chief: H. Peters (Maastricht University, The Netherlands); *Honorary Editor*: S.H. Tijs (Tilburg University, The Netherlands).
Editorial Board: E.E.C. van Damme (Tilburg University, The Netherlands); H. Keiding (University of Copenhagen, Denmark); J.-F. Mertens (Université catholique de Louvain, Belgium); H. Moulin (Rice University, Houston, USA); Shigeo Muto (Tokyo University, Japan); T. Parthasarathy (Indian Statistical Institute, New Delhi, India); B. Peleg (Hebrew University, Jerusalem, Israel); T.E.S. Raghavan (University of Illinois at Chicago, USA); J. Rosenmüller (University of Bielefeld, Germany); A. Roth (Harvard University, USA); D. Schmeidler (Tel-Aviv University, Israel); R. Selten (University of Bonn, Germany); W. Thomson (University of Rochester, USA).

Scope: Particular attention is paid in this series to game theory and operations research, their formal aspects and their applications to economic, political and social sciences as well as to socio-biology. It will encourage high standards in the application of game-theoretical methods to individual and social decision making.

The titles published in this series are listed at the end of this volume.

Theo S. H. Driessen · Gerard van der Laan
Valeri A. Vasil'ev · Elena B. Yanovskaya
Editors

Russian Contributions to Game Theory and Equilibrium Theory

Springer

Theo S. H. Driessen
University of Twente
Faculty of Electric Engineering,
Mathematics, and Computer Science
Drienerlolaan 5
7522 NB Enschede
The Netherlands
E-mail: t.s.h.driessen@ewi.utwente.nl

Gerard van der Laan
Vrije Universiteit, Amsterdam
Faculty of Economics
and Business Administration
De Boelelaan 1105
1081 HV Amsterdam
The Netherlands
E-mail: glaan@feweb.vu.nl

Valeri A. Vasil'ev
Russian Academy of Sciences, Novosibirsk
Sobolev Institute of Mathematics
Prosp. Koptyuga, 4
630090 Novosibirsk
Russia
E-mail: vasilev@math.nsc.ru

Elena B. Yanovskaya
Russian Academy of Sciences, St. Petersburg
St. Petersburg Institute for Economics
and Mathematics
Tchaikovsky st., 1
191187 St. Petersburg
Russia
E-mail: eyanov@iatp20.spb.org

This book is a collection of journal articles translated from Russian.

Cataloging-in-Publication Data
Library of Congress Control Number: 2006920772

ISBN-10 3-540-31405-9 Springer Berlin Heidelberg New York
ISBN-13 978-3-540-31405-9 Springer Berlin Heidelberg New York

This work is subject to copyright. All rights are reserved, whether the whole or part of the material is concerned, specifically the rights of translation, reprinting, reuse of illustrations, recitation, broadcasting, reproduction on microfilm or in any other way, and storage in data banks. Duplication of this publication or parts thereof is permitted only under the provisions of the German Copyright Law of September 9, 1965, in its current version, and permission for use must always be obtained from Springer-Verlag. Violations are liable for prosecution under the German Copyright Law.

Springer is a part of Springer Science+Business Media
springeronline.com

© Springer-Verlag Berlin Heidelberg 2006
Printed in Germany

The use of general descriptive names, registered names, trademarks, etc. in this publication does not imply, even in the absence of a specific statement, that such names are exempt from the relevant protective laws and regulations and therefore free for general use.

Cover design: Erich Kirchner
Production: Helmut Petri
Printing: Strauss Offsetdruck

SPIN 11532279 Printed on acid-free paper – 43/3153 – 5 4 3 2 1 0

Preface and acknowledgments

The purpose of this collection of papers is to report about highly qualified research by Soviet game theorists during the two decades 1968-1988 to the international readership. The research by Soviet scientists within the field of game theory, starting from the early period around 1965 under the supervision of their former leader N.N. Vorob'ev, has resulted in many high-level contributions. The contributions in this collection have not been published before, except in Russian language journals. In the past the Russian literature in general, and the Russian game theoretical literature in particular, was not available for colleagues outside the Soviet Union, and even nowadays the papers in the Russian language are inaccessible for the international readership. Therefore this collection of English translations of specifically chosen former research articles in Russian will help to close the gap in the international knowledge about the Soviet advances in game theory.

In 1968, the Econometric Research Program at Princeton University published a bundle of English translations of Soviet research, entitled "Selected Russian Papers on Game Theory, 1959-1965". This bundle, yet recently downloadable from www.econ.princeton.edu/ERParchives, played an important role to inform the international game theory community about interesting game theoretic results in the former Soviet Union. In particular, because of this bundle the international researchers in game theory learned that the famous existence result for the core had been proven by the Russian PhD student Olga Bondareva already five years before the first paper by Shapley on this topic was published.

Since the 1990s also the Russian researchers are publishing in English and therefore their results are available to the international readership. However, the period 1965-1990 remains as a period in which within the former Soviet Union game theory was successfully developing, but separately from the world science. This collection of papers attempts to acquaint English language readers with some contributions in game theory and the related field of equilibrium theory, which never had been published in English before. Since some Soviet journals of high level were translated into English, the most important contributions of the Soviet researchers to these fields are already accessible in English. For this reason there are no papers of the Soviet game-theoretic leader N.N.Vorob'ev within this volume. Nevertheless, many papers containing very nice results are still unattainable for foreign readers. The twelve selected papers in this volume, all from the period 1965-1990, have been translated by the authors themselves. Some of the papers are slightly adapted to improve the readability. In addition, the volume starts with an introductory chapter on the history of Soviet game theory before the 1990s. This chapter also contains a short summary of the selected contributions.

The idea of editing a volume of English translations of former Soviet contributions emerged from the continuous scientific communication between game theorists from The Netherlands and The Russian Federation during the last decade 1996-2005. The mutual scientific communication was initiated in June 1996 by the participation of Dutch researchers into the international conference on game theory in memoriam of N.N. Vorob'ev, held at St. Petersburg. This conference initiated subsequent scientific visits by Russian game theorists to The Netherlands, in particular the University of Twente at Enschede. These visits, realized through funding from the Dutch mathematical organization SWON (Stichting Wiskunde Onderzoek Nederland), resulted in several joint publications. In 2001 the individual SWON funding was upgraded by the Dutch scientific organization NWO (Nederlandse Organisatie voor Wetenschappelijk Ondezoek) within the framework of a bilateral scientific agreement between The Netherlands and The Russian Federation. We are very grateful to NWO and the Russian Foundation for Basic Research (RFBR) for the financial support for our projects "Axiomatic Approach to the Elaboration of Cooperative Game Solutions: Theory and Applications" (2001-2005) and "Game Theoretic Models for Cooperative Decision Making and Their Applications to Mathematical Modelling in Economics and Social Sciences" (2005-2008). Within these projects researchers participate from University of Twente at Enschede, Vrije Universiteit at Amsterdam, Tilburg University, St. Petersburg Institute for Economics and Mathematics, Sobolev Institute of Mathematics at Novosibirsk and Central Institute of Economics and Mathematics at Moscow. As part of these projects three international three-days workshops on cooperative game theory have been organized in Enschede (2002, 2005) and Amsterdam (2004). From the ongoing cooperation and mutual discussions the Dutch editors learned that the Russian literature on game theory and related fields contains many contributions which deserve to become accessible for the international readership. We owe many thanks to the authors for their efforts to translate their old papers in English. The current volume will contribute to a better scientific knowledge of the treasury hidden in the former Soviet journals.

We conclude to express our thanks to Hans Peters from Maastricht University for his invaluable help as managing editor of this volume and his secretary Yolanda Paulissen for processing the papers into their final form.

Theo Driessen — Enschede
Gerard van der Laan — Amsterdam
Valeri Vasil'ev — Novosibirsk
Elena Yanovskaya — St. Petersburg

December 2005

Contents

Chapter 1

Introduction

V.A. Vasil'ev and E.B. Yanovskaya

1 Game theory in the USSR before 1990

In this introductory chapter first a short historical information about the development of game theory in the USSR before the 1990s is given. It should help to understand the choice of the papers in this volume. The development of game theory in the USSR began in the early sixties. N.N.Vorob'ev (1925-1995) was the author of the first papers about game theory in Russian; he was also the leader of a group of PhD students and young researchers at Leningrad who studied this new field in mathematics.

The early papers by Vorob'ev on enumerating equilibrium points in bimatrix games ([74]), coalitional games where a player may belong to different coalitions ([77] and [78]) and equivalence of different types of strategies in games in extensive form ([75] and [76]), were published in Soviet journals having English translations, and so they became known in the West. Later on, game theory in the USSR exhibited significant progress, which deserved world-wide attention. However, many game-theoretic papers were published in Soviet journals and edited volumes which have never been translated into English.

In the 1960s, the theory of zero-sum games was popular in the USSR. Some mathematicians even considered minmax theorems as the only truly mathematical results in game theory. New minmax theorems were obtained at that time, see e.g. Yanovskaya [82] and [83]. Also the structure of the set of optimal strategies was studied and for some particular classes of zero-sum games optimal strategies were found. Similar results for general noncooperative games were obtained as well. The axiomatizations of the value of a matrix game and equilibrium payoffs in non-cooperative games were obtained by the Lithuanian researcher Vilkas [72], who was one of the first PhD students of Vorob'ev. A review of all these results can be found in the comments chapters to Vorob'ev's book "Foundations of Game Theory: Non-cooperative Games" [80].

One of the main problems of the theory of games in extensive form is to find the classes of the most simple mixed strategies which are sufficient for determining the optimal ones. This field of research began with the famous Kuhn theorem in which he proved for extensive form games with perfect recall the equivalence of each mixed strategy of a player to some behavioural strategy. This theorem has been generalized by several researchers in different directions. In 1957 Vorob'ev published his first paper [75] within a series of generalizations of Kuhn's equivalence theorem about behavioral strategies in extensive form games with perfect recall. He considered more complicated structures of players' recall. For the case of "ordered" recall he introduced "reduced" strategies which were extensions of behavioral strategies and he proved the corresponding equivalence theorem. In [76], Vorob'ev considered another structure of recall and proved an equivalence theorem generalizing that of Thompson for signalling strategies.

Romanovsky [61] found a method of reducing the problem of finding optimal behavioural strategies in finite two-person games in extensive form with perfect recall to the solution of matrix games with linear restrictions on the set of their mixed strategies. For games with partially ordered recall Yanovskaya [81] defined "quasi-strategies" described by the probabilities of vertices and alternatives of the information structure of each player. She proved the corresponding equivalence theorem for quasi-strategies and reduced optimal quasi-strategies of a two-person game in extensive form to the solution of a polyhedral game. Konurbaeva [40] extended this result to games described by acyclic graphs instead of trees. In [41], she also considered games in extensive form where (as Isbell defined) information sets may precede themselves. For such games she defined special "linear-like strategies" and proved the corresponding equivalence theorem. At last, recently Liapunov [49] introduced "H-strategies" for games in extensive form maximizing the entropy of mixed strategy in the class of all strategies equivalent to a given one. The corresponding equivalence theorem has been proved. For games with perfect recall the H-strategies coincide with the behavioural ones.

Another direction in game theory and decision making under uncertainty that has been developed in the 1970s at Leningrad concerns mixed extensions of arbitrary binary relations. Some papers on this topic have been published and the results are summed up in Kiruta *et al.* [39]. These results are about the existence of maximal elements of mixed extensions of binary relations and their applications to the existence of mixed equilibrium points in non-cooperative games with ordered outcomes as well as some mixed concepts in cooperative TU-games. The bibliography of [39] shows that the well-known skew-symmetric utility theory due to Fishburn [11] has also been developed at the same time in the former Soviet Union (see also the papers of Yanovskaya in this volume).

The theory of cooperative games was developing in Leningrad beginning from the famous paper due to Bondareva [25] (see also [26]) who obtained necessary and sufficient conditions for non-emptiness of the core of a TU-game. Later on sufficient conditions providing stability of the core were obtained by Diubin (see the paper in this volume), Bondareva [27], Kulakovskaya [44] and Vilkov [73] for NTU-games, and necessary and sufficient conditions by Kulakovskaya [43]. For cooperative games with a countable player set, Naumova [53], [54] and [55] respectively, obtained necessary and sufficient conditions for non-emptiness of the countably-additive core, found the NM-solutions for some classes of simple games and she obtained sufficient conditions for the existence of M_i^1 bargaining sets.

All the necessary and sufficient stability conditions, mentioned above, indicate that the core of a TU-game may be quite often unstable with respect to the classic domination relation. The same is true even for the standard transitive closure of this relation. Nevertheless, as it was established by Vasil'ev (see his paper in this volume), the core is externally stable with respect to the so-called limit transitive closure, being a fairly natural "sequential" transitive extension of the classic domination relation. An example of an NTU-game having a closed subset of imputations, which contains all the monotonic trajectories originating at its points and does not intersect with the core, demonstrates that the side-payment property is of crucial importance for this result (for more details, see the same paper).

The most important achievements of Soviet game theorists regarding to the classic VonNeumann-Morgenstern solution, is the well-known existence theorem established for all different cases in four-person TU-games in Bondareva [28], Kulakovskaya [44], [45], [46] and Bondareva *et al.* [29]. Also Vasil'ev's later obtained results in this field on the existence and NM-rank of the so-called generalized NM-solutions, clarifying the structure of the classic domination relation, are of interest (see [69], [70], and Vasil'ev's paper in this volume).

In the mid of seventies a new approach for studying the Shapley value and its nonsymmetric analogues, both in finite and infinite case, has been elaborated by Vasil'ev at Novosibirsk. This approach strongly relied upon the vector lattice theory in order to provide a systematic treatment of the values by applying the advanced methods of Kantorovich-Banach-space theory [37], [38] (for the basic notions see, e.g., [1]).[1] The important concept of a totally positive game has been introduced, and a new axiomatization of the Shapley value, treated as a positive (w.r.t. the cone of totally positive games) symmetric linear operator, has been given in [66]. By applying this axiomatization, so-called Harsanyi payoff vectors and the Harsanyi set (also known as the Selectope, independently introduced in Hammer *et al.* [14]), were proposed by Vasil'ev for to give a unified description of the core imputations in terms of some decentralization mechanisms [67]. Surprisingly, the Harsanyi set turned out to be of a core-type structure itself, a fundamental result first established by Vasil'ev in 1980 [68] (see also the paper in this volume), and has been independently rediscovered by Derks *et al.* [8].

To conclude this survey on the classic cooperative game theory results, obtained in Leningrad and Novosibirsk during 1968-1990, we mention that in 1975 Sobolev [65] published a very important result characterizing the prenucleolus axiomatically, and among the axioms the consistency axiom (or the reduced game property) was used for the first time. The very complicated proof can be found in Peleg and Sudhölter's introductory book [21] on cooperative game theory. In [65] also an alternative consistency property is defined and used to obtain a new axiomatization of the Shapley value.

Moscow game theory was initiated in the late 1960s and early 1970s by Germeier (1918-1975), head of the Operations Research Departments of Moscow State University and the Computing Center of the USSR (now Russian) Academy of Sciences. The most popular topic was "hierarchic games", close in spirit to the principal-agent problems: when the principal agent lacks information, he is assumed to follow the maxmin approach rather than form subjective beliefs. Typical settings and solutions are presented in [33]. Unfortunately, even the original book was published posthumously, and its editing was not very good. An interesting model was developed by Germeier and Vatel' [34], who considered voluntary contributions to collective goods. Later on Gurvich [35] and [36], former PhD student of Germeier, characterized normal forms of games with perfect information, respectively

[1] Interestingly to note that both vector lattice theory, adapted to the description of the Shapley value, and linear programming, applied to the core existence problem, were created, among others, by the Soviet academician and Nobel prize winner L.V.Kantorovich, former vice-director of the Institute of Mathematics, Siberian Branch of the USSR Academy of Sciences.

two person game forms ensuring the existence of a Nash equilibrium.

Both in Moscow and in Leningrad bargaining solutions were investigated almost parallel to analogous researches in the west. In 1976 Butrim [31], pupil of Germeier, independently of Kalai and Smorodinsky [16], defined a system of axioms for two-person bargaining problems leading to the Kalai-Smorodinsky solution. Further, Butrim [32] extended the result to n-person bargaining problems. Vorob'ev [79] characterized the n-person Nash bargaining solution without the assumption about the existence of a feasible payoff vector strongly greater then the status quo. Independently of Myerson [19], Perles and Maschler [22] and Thomson [24], Pechersky studied the linear and superlinear bargaining solution (see his two papers in this volume).

In the 1960s, systematic research in the theory of differential games was started by the Soviet mathematicians to supply the scientific basement and rigorous techniques for the solution of important practical problems (such as pursuit-evasion problem, control under uncertainty, search problems and others). Initially main attention was paid to the zero-sum differential games. During the 1960-1970s the Soviet academicians Krasovsky [42] and Pontryagin [59], [60] (and the scientific schools of these mathematicians, see also Petrosjan [57] created a quite general and rigorous theory of so-called positional differential games.

During the 1970-1980s the general theory of n-person (non-cooperative and cooperative) differential games began to develop. In particular, Petrosjan [56] proposed the concept of time-consistency of differential n-person game solutions as a fundamental property of stability. This concept is actively developed at St. Petersburg University. The references can be found in [58].

In the middle of the seventies, working on a general conception of game theory as a mathematical discipline, Vorob'ev, impressed by a paper by Bubelis [30] on the algebraic reduction of n-person games to 3-person games and some results by Yanovskaya concerning equilibria in the mixed extension of general non-cooperative games (in this volume), formulated an idea of categorical approach to game theory. According to his original intuition, non-cooperative games form a category, and the set of equilibrium points (and, perhaps, some other solution concepts) is a functor from this category. Although the realization of this program turned out to be more complicated than was perceived by Vorob'ev, the problem was successfully resolved by Lapitsky (see his paper in this book and his later contributions [47] and [48] in Russian), who constructed a general theory of categories of games and corresponding solution concepts as properly chosen functors.

It is common knowledge that game theory methods and models are the corner-stones of modern general equilibrium theory in economics (see e.g. [20], [23] and [15]). We just mention the famous core equivalence results by Aumann [2], [3] and Aumann-Shapley [5], linking together the Walrasian equilibria, the core and the Shapley value in nonatomic pure exchange economies. Vice versa, it is a well-known fact that the core and Nash equilibrium, the most popular and highly elaborated game theory concepts, can be traced back to the Edgeworth contract curve [10] and the Cournot equilibrium [7], respectively (not speaking much about many other notions designed by the economists about one-two hundreds years ago before these notions were properly incorporated into the game theory thesaurus). Therefore we would pay some attention in this book to equilibria in economic models. Unlike the pure game theory papers, most of the main results obtained by Soviet researchers during the 1960-1980s concerning applications of game theory to the economical equilibrium analysis, were either published directly in the international journals (e.g., Econometrica, Journal of Mathematical Economics, Mathematical Social Sciences and others), or are translated in English somewhere, mostly in Matekon (M.E.SharpInc., Armonk, NY 10504), a well-known journal that helpfully served to integrate many Soviet mathematical economists into the international academic community (see e.g. [52]). That is why we restrict ourselves by presenting merely a small number of Russian papers in mathematical economics that seem to be of rather strong interest till now, but has not been translated in English yet.

Since a brief survey of the papers on mathematical economics included in the book is given in the section below, we conclude with some core equivalence results, obtained in the former USSR for the Lindahlian type economies by Vasil'ev [71] and Makarov and Vasil'ev [51]. In their papers (in Russian and hardly available to the Western readers), seminal concepts of blocking in replicas of the pure public goods economies and economies with externalities were proposed and the validity of the Edgeworth conjecture on the shrinkability of the cores to the Lindahlian equilibria and so-called informational equilibria, respectively, has been proven (for more details, see also [52]).

2 Summary of the selected contributions

The twelve papers in this book can be subdivided into four groups: noncooperative game theory and social choice, cooperative game theory, bargaining theory and equilibrium theory. The two papers of Yanovskaya fit within the first group. In both papers mixed extensions of arbitrary binary relations are used as the key tool. Till the 1970s the probabilistic approach to decision-making was based mainly on expected utility which is applicable only for transitive preferences. Mixed extensions of non-transitive binary relations were begun to be investigated with the help of skew-symmetric bilinear theory of expected utility, introduced independently by Kiruta *et al.* [39] and Fishburn [11]. This theory was applied in both papers. In the first paper one probabilistic social choice rule on a finite set of alternatives is defined and an axiomatic characterization of the rule is given. It is shown that the set of maximal elements of this rule coincides with the set of all optimal mixed strategies of every player in the symmetric matrix game whose entries are the differences between the number of individuals prefering the row alternative above the column alternative and the number of individuals having the inverse opinions. In the second paper more general results on non-cooperative games with ordered and arbitrary outcomes are presented. In particular, theorems establishing nonemptiness of the sets of mixed equilibrium points for non cooperative games with ordered outcomes and also with non-transitive outcomes are proved. In the paper of Lapitsky a general theory of categories of games is developed and corresponding "equilibrium-type" solution concepts as properly chosen functors are given, see also the remarks on this topic in the previous section.

The second group of papers (about cooperative game) theory consists of three papers, that is two short papers of Diubin and one long paper of Vasil'ev. In the first paper of Diubin the results by Gillies [13] and Bondareva [27] on the stability of the core are strengthened. It is shown that the upper estimates of the characteristic function values in the class of symmetric estimates are the best possible. In the second paper it is shown that for cooperative games with a countable set of players the Shapley value can be determined as the unique measure on the set of players on which the minimum of some quadratic functional is attained. This is a generalization of a result of Keane [17] for finite TU-games.

Vasil'ev's paper is mostly devoted to the comparative analysis of some classes of nonsymmetric values, core allocations, generalized von Neumann-Morgenstern solutions and totally-contractual sets. It contains an extended survey of results obtained by the author on the problems in question and consists of three parts. In the first part a unified functional approach to the investigation of nonsymmetric analogues of the Shapley value, already mentioned in the previous section, is given. In the second part the so-called generalized VonNeumann-Morgenstern solutions (gNM-solutions, shortly) are introduced, based on the principle of sequential improvements of dominated alternatives. The notion of NM-rank, characterizing the number of improvements required to arrive at gNM-solution starting at the "most distant point", is proposed, and some gNM-existence theorems with evaluation of NM-rank are established. It is shown, that in contrast to the classic NM-solution, generalized NM-solutions always exist. Within the framework of a topological approach, inspired by the well-known Maschler-Peleg theory of dynamic systems [18], it is proven that for any balanced cooperative TU-game there exists a dynamic system, whose final set is globally stable and equals the core, while outside the core all the outcomes of the transition function dominate the current imputation. As a consequence one of the main results of the paper follows: for any cooperative TU-game there exists an NM-solution w.r.t. the limit transitive closure of the classic domination relation. This solution is unique and coincides with the core, if the latter is not empty, and can be chosen to contain a finite number of imputations otherwise. The third, concluding part of the paper contains a game-theoretical analysis of the totally-contractual sets, similar to that introduced by Makarov [50] in order to describe stable outcomes of some rather natural recontracting procedure in pure exchange economies. The structure of domination relations, induced by several rules of entering and breaking contract systems, is studied, and quite natural and mild conditions, providing the coincidence of the totally-contractual core and Walrasian equilibria, are established.

The third group of papers about bargaining theory consists of two papers by Pechersky and one paper by Kukushkin. In the first paper of Pechersky the linear bargaining solutions are characterized by efficiency and linearity. By that time such solutions had already been characterized by Myerson [19] and Thomson [24]. However, unlike these papers, Pechersky's paper does not impose any restriction (such as strict convexity or smoothness) on the bargaining sets. The solution is defined as the Steiner point of a contact set of feasible outcomes and the supporting hyperplane with a unit normal vector. In the second paper of Pechersky the superlinearity axiom is used for the first time and the existence of the superlinear solution for n-person bargaining games satisfying continuity, symmetry, weak Pareto optimality, superlinearity, and translation covariance, was proved. Later Perles and Maschler [22] defined and characterized the super-additive solution for two person bargaining games satisfying Pareto optimality, scale transformation covariance, super-additivity, symmetry and continuity. However, their solution cannot be extended to n-person bargaining games for $n > 2$. The paper of Kukushkin considers the following problem: two agents derive transferable utility from alternatives in a feasible set, but the choice is made by a third agent, who is only concerned in payments from the interested agents. The concept of equilibrium is given and it is shown that there exists an equilibrium that Pareto dominates (in the viewpoint of the two interested agents) any other equilibrium. In the terminology of Bernheim and Whinston [6] it is a menu auction.

The concept of equilibrium plays a key role in game theory. The fourth and last group of papers is on related equilibrium concepts in economic theory and consists of three papers by Zak, Shmyrev, and Danilov and Sotskov. The paper of Zak may seem to be rather technical, but is very interesting from both a theoretical and applied point of view. A differential approach is proposed to develop an original systematic consideration of stable and unstable economies. In the sense of Aumann and Peleg [4], a pure exchange economy is unstable when one of the participants can improve its position in equilibrium by throwing out, or hiding, a part of its initial endowment (see also [12] for a coalitional type of this phenomenon, which is less significant in competitive environments). The paper contains one of the main results in the field, establishing the stability of Walrasian economies with normal demand and equilibrium gross substitutability. It is also worth to note that an advanced differential topology technique, applied and partially developed by the author to study individual strategy-proofness aspects of Walrasian equilibrium, may be of particular importance to extend some well-known (at least in Russia) results of Polterovich and Spivak [63] on coalitional stability, as well. Shmyrev's paper deals with a new approach for searching Walrasian equilibrium in a linear pure exchange economy with fixed incomes. The approach is based on the consideration of a special linear parametric transportation problem, with prices to be taken as parameters. As it was shown in a later article [64], in comparison with the first well-known finite algorithm by Eaves [9], Shmyrev's finite methods for searching equilibrium, being elaborated by applying both complementarity theory and rather efficient linear programming-type procedures, seem to be more practical, even for the most general linear exchange models. Finally, the paper of Danilov and Sotskov concerns some analogies of the so-called equilibrated states as introduced by Polterovich [62]. A generalization of the famous Bergstrom-Ky-Fan theorem on maximal elements is given, and a unified approach to the existence theorems for both equilibrated states and the cores of cooperative games, based on this generalization, is proposed. Besides the cooperative game theory, the main results may also be of interest in some problems of optimal allocation of resources under rigid prices.

References

1. Aliprantis, C.D., and K.C. Border, *Infinite Dimentional Analysis: a Hitchhiker's Guide*, Springer-Verlag, Berlin, 1994.
2. Aumann, R.J., Markets with a continuum of traders, *Econometrica* 32, 1964, 39–50.
3. Aumann, R.J., Values of markets with a continuum of traders, *Econometrica* 43, 1975, 611–646.
4. Aumann, R.J., and B.Peleg, A note on Gale's example, *J. of Math. Econ.* 1, 1974, 209–211.
5. Aumann, R.J., and L.S. Shapley, *Values of Non-Atomic Games*, Princeton University Press, Princeton NJ, 1974.
6. Bernheim, B.D., and M.D. Whinston, Menu auctions, resource allocation, and economic influence, *Quarterly J. of Economics* 101, 1986, 1–31.
7. Cournot, A., *Recherches sur les Principes Mathematiques de la Theorie des Rechesses*, Hachette, Paris, 1838.
8. Derks, J., H. Haller and H. Peters, The Selectope for cooperative games, *Intern. J. Game Theory* 29, 2000, 23–38.
9. Eaves, B.S., A finite algorithm for the linear exchange model, *J. of Math. Econ.* 3, 1976, 197–204.
10. Edgeworth, F.Y., *Mathematical Psysics*, Kegan Paul, London, 1881.
11. Fishburn, P., Non-cooperative stochastic dominance games, *Intern. J. Game Theory* 7, 1978, 51–61.
12. Gale, D., Exchange equilibrium and coalitions: an example. *J. of Math. Econ.* 1, 1974, 63– 66.
13. Gillies, D.B., Some solutions to general non-zero-sul games, in: *Contributions to the Theory of Games* 4, A.W.Tucker and R.D.Luce, eds., Princeton University Press, Princeton NJ, 1959, 47–85.
14. Hammer, P.L., U.N. Peled and S. Sorensen, Pseudo-boolean functions and game theory, I: Core elements and Shapley value, *Cah. Cent. Etud. Rech. Oper.*, 19, 1977, 159–176.

15. Hildenbrand, W., Core of an economy, in: *Handbook of Mathematical Economics* 2, K.J. Arrow and M.D. Intrilligator, eds., North-Holland, Amsterdam-New York-Oxford, 1982, 831– 877.

16. Kalai, E., and M. Smorodinsky, Other solution to Nash's bargaining problem, *Econometrica* 43, 1975, 513–518.

17. Keane, M., *Some Topics in N-person Game Theory*, Thesis, Northwestern University, Evanston, Illinois.

18. Maschler, M., and B. Peleg, Stable sets and stable points of set-valued dynamic systems with applications to game theory, *SIAM J. Control Optim.* 14, 1976, 985–995.

19. Myerson, R., Utilitarianism, egalitarianism and the timing effect in social choice problem, *Econometrica* 49, 1981, 883–897.

20. Neumann, J. von, and O. Morgenstern, *Theory of Games and Economic Behavior*, Princeton University Press, Princeton, NJ, 1944.

21. Peleg, B., and P. Sudhölter, *Introduction to the Theory of Cooperative Games*, Kluwer Acad. Publishers, Boston, 2003.

22. Perles, M., and M. Maschler, The super-additive solution for the Nash bargaining game, *Intern. J. Game Theory* 10, 1981, 163–193.

23. Shubik, M., Game theory in political economy, in: *Handbook of Mathematical Economics* 1, K.J. Arrow and M.D. Intrilligator, eds., North-Holland, Amsterdam-New York-Oxford, 1981, 285–330.

24. Thomson, W., Nash's Bargaining solution and utilitarian choice rules, *Econometrica* 49, 1981, 535 – 538.

Russian papers

25. Bondareva O.N., The theory of the core of an n-person game, *Vestnik Leningrad. Univ.* 13, 1962, 141–142 (in Russian), also in: *Selected Russian Papers on Game Theory, 1959–1965*, Economic Research Program, Princeton University, 1968, 29–31.

26. Bondareva O.N., Several applications of linear programming methods to the theory of cooperative games, *Problemi Kibernetiki*, 1963 (N10), 119–139 (in Russian). English translation in: *Selected Russian Papers on Game Theory, 1959–1965*, Economic Research Program, Princeton University, 1968, 79–114.

27. Bondareva O.N., The existence of a solution, coinciding with the core, in an n-person game, in: *Proceedings of the VI Soviet Union's Conference on Theory of Probabilities and Mathematical Statistics (Vilnius, 1960)*, Vilnius, 1962, 337 (in Russian). English translation in: *Selected Russian Papers on Game Theory, 1959-1965*, Economic Research Program, Princeton University, 1968, 69.

28. Bondareva O.N., The solution of a classical cooperative four-person game with a nonempty core (general case), *Vestnik Leningrad Univ. (Matematika)*, 1979 (N19), 14–19 (in Russian). English translation: *Vestnik Leningrad Univ. Math.* 1980 (N12), 247–253.

29. Bondareva O.N., T.E. Kulakovskaya and N.I. Naumova, Solution of arbitrary four-person cooperative game, *Vestnik Leningrad. Univ. (Matematika)*, 1979 (N7), 104–105 (in Russian). English translation: *Vestnik Leningrad. Univ. Math.* 1979, 131–139.

30. Bubelis V., On equilibria in finite games, *Intern. J. Game Theory* 8, 1979, 65–79.

31. Butrim B.I., A modified bargaining solution, *Journal Vychislitelnoj Matematiki i Matematicheskoj Physiki* 16, 1976, 340–350 (in Russian).

32. Butrim B.I., N-person games with essential set of criteria, *Journal Vychislitelnoj Matematiki i Matematicheskoj Physiki* 18, 1978, 62–72 (in Russian).

33. Germeier, Yu.B., *Non-Antagonistic Games*, 1976. (Translated from Russian by A.Rapoport, D.Reidel Publishing Company, Dordrecht, 1986.)

34. Germeier, Yu.B., and I.A. Vatel', On games with a hierarchical vector of interests. *Izvestiya Akademii Nauk SSSR. Tekhnicheskaya Kibernetika*, 1974 (N3), 54–69 (in Russian). English translation: *Optimization Techniques*, 1974, 460–465.

35. Gurvich, V.A., On the normal form of positional games. *Doklady Akademii Nauk SSSR* 264, 1982, 30–33 (in Russian). English translation: *Soviet Mathematics. Doklady* 25, 1982, 572–575.

36. Gurvich, V.A., Equilibrium in pure strategies, *Doklady Akademii Nauk SSSR*, 303, 1988 789–793 (in Russian). English translation: *Soviet Mathematics. Doklady* 38, 1989, 597–602.

37. Kantorovich L.V., On partially ordered vector spaces and their applications to the theory of linear operators, *Doklady Akademii Nauk SSSR*, 4, 1935, 11–14 (in Russian).

38. Kantorovich L.V., *Selected Works. Part I: Descriptive Theory of Sets and Functions. Functional Analysis in Semi-ordered Spaces*, (Classics of Soviet Mathematics Series, ISSN 0743-9199, Vol.3), Gordon and Breach Publishers, Amsterdam, 1996.

39. Kiruta A.Ya., A.M. Rubinov and E.B. Yanovskaya, *Optimal Choice of Distributions in Complex Socio-Economical Problems*, Nauka, Leningrad, 1980 (in Russian).

40. Konurbaeva B.M., Quasi-strategies in games on graphs, *Optimizatsija* 2, 1971, 74–89 (in Russian).

41. Konurbaeva B.M., Linear-like strategies in Isbell's games, in: *Advances in Game Theory, Proc. of the Second USSR Game Theory Conference (Vilnius, 1971)*, E.Vilkas, ed., Mintis, Vilnius, 1973, 177–180 (in Russian).

42. Krasovsky N.N., and A.I. Subbotin, *Positional Differential Games*, Nauka, Moscow, 1974, 456p.

43. Kulakovskaya T.E., Necessary and sufficient conditions for coincidence of the core with the NM solution in TU games, *Doklady Akademii Nauk SSSR* 199, 1971, 1015–1017 (in Russian). English translation: *Soviet Mathematics. Doklady* 12, 1971, 1231–1234.

44. Kulakovskaya T.E., Sufficient conditions for coincidence of the core with the NM solution, *Litovsky Matematichesky Sbornik* 9, 1969, 424–425 (in Russian).

45. Kulakovskaya, T.E., The solution of a class of cooperative four-person games with non-empty core, *Vestnik Leningrad. Univ. (Matematika)*,

1979 (N19), 42–47 (in Russian). English translation: *Vestnik Leningrad. Univ. Math.* 1980 (N12), 247–253.

46. Kulakovskaya, T.E., NM-solutions of some cooperative four-person games with empty core, *Vestnik Leningrad. Univ. (Matematika)*, 1979 (N7), 52–60 (in Russian), English Translation: *Vestnik Leningrad. Univ. Math.*, 1980 (N12), 301–314.

47. Lapitsky V., On axiomatics of equilibrium I, II, *Mathematical Methods in Social Sciences (Vilnius)* 15, 1982, 18 – 26 and *Mathematical Methods in Social Sciences, (Vilnius)* 17, 1984, 18 – 23 (in Russian).

48. Lapitsky V., Categories of games with fixed set of players, *Mathematical Methods in Social Sciences (Vilnius)* 16, 1983, 55 – 76 (in Russian).

49. Liapunov A.N., H-strategies in extensive form games, *International Game Theory Review* 1, 1999, 273–281.

50. Makarov V.L., Economical equilibrium: existence and extremal properties, *Problems of the Modern Mathematics* 19, 1982, 23–59 (in Russian).

51. Makarov V.L., and V.A. Vasil'ev, Information equilibrium and the core in generalized exchange models, *Doklady Akademii Nauk SSSR* 275, 1984, 549–553 (in Russian). English translation: *Soviet Math. Dokl.* 29, 1984, 264–268.

52. Makarov V.L., and V.A. Vasil'ev, Equilibrium, rationing and stability, *Matekon* 25, 1989, 4–95.

53. Naumova N., On the core in games with countable number of players, *Doklady Akademii Nauk SSSR* 197, 1971, 40–42 (in Russian). English translation: *Soviet Math. Dokl.* 12, 1971, 409–411.

54. Naumova N., Solutions of infinite simple games. *Operations Research and Statistical Modelling* 1, 1972, 126–135 (in Russian).

55. Naumova N., Sufficient conditions for existence of M_i^1-bargaining sets, in: *Advances in Game Theory, Proc. of the Second USSR Game Theory Conference (Vilnius, 1971)*, E.Vilkas, ed., Mintis, Vilnius, 1973, 146–149 (in Russian).

56. Petrosjan L.A., Solutions stability in differential n-person games, *Vestnik Leningrad. Univ.*, 1977 (N19), 46–52 (in Russian).

57. Petrosjan L.A. *Differential Games of Pursuit*, Series on Optimization 2, World Scientific, Singapore, 1993, 325p.

58. Petrosjan L.A., and N.A. Zenkevisn, *Game Theory*, Series on Optimization 3, World Scientific, Singapore, 1996, 352p.

59. Pontryagin L.S., On linear differential games I, *Doklady Akademii Nauk SSSR* 174, 1967 (N6) (in Russian).

60. Pontryagin L.S., On linear differential games II, *Doklady Akademii Nauk SSSR* 175, 1967 (N4) (in Russian).

61. Romanovsky J.V., On reducing a game with perfect recall to a matrix game. *Doklady Akademii Nauk* 144, 1962, 62–64 (in Russian).

62. Polterovich V.M., Equilibrated states and optimal allocations of resources under rigid prices, *J. Math. Econ.* 19, 1990, 255–268.

63. Polterovich V.M., and V.A. Spivak, Coalitional Stability of Economic Equilibrium, CEMI AN SSSR, Moscow, 1978 (in Russian).

64. Shmyrev V.I., An algorithm of searching for the equilibrium in linear exchange model, *Sibirsk. Mathem. Journal* 26, 1985, 162–175 (in Russian).

65. Sobolev, A.I., (1975), The characterization of optimality principles in cooperative games by functional equations, in: *Mathematical Methods in the Social Sciences (Vilnius)* 6, 1975, 94–151 (in Russian).

66. Vasil'ev V.A., The Shapley value for cooperative games of bounded polynomial variation, *Optimizacija* 17, 1975, 5–26 (in Russian).

67. Vasil'ev V.A., Support function of the core of a convex games, *Optimizacija* 21, 1978, 30–35 (in Russian).

68. Vasil'ev V.A., On the H-payoff vectors for cooperative games, *Optimizacija* 24, 1980, 18–32 (in Russian).

69. Vasil'ev V.A., Lucas game has no NM-solution in H-imputations, *Optimizacija*, 27, 1981, 5–20 (in Russian).

70. Vasil'ev V.A., *Exchange Economies and Cooperative Games*, Novosibirsk State Univ. Press, Novosibirsk, 1984 (in Russian).

71. Vasil'ev V.A., Core asymptotics in the Lindahlian-type models, *Optimizacija* 41, 1987, 15–35 (in Russian).

72. Vilkas E., An axiomatic characterization of the value of matrix games, *Teorija verojatnostei i ee primenenia* 8, 1963, 324–327 (in Russian). English translation: *Theory of Probability and its Applications* 8, 1963, 304–307.

73. Vilkov V.B., Some theorems on a stable core for games without side payments, *Vestnik Leningrad. Univ.*, 1972 (N19) and *Ser. Mat. Mekh. Astron.* (N1), 5–8 (in Russian).

74. Vorob'ev N.N., Equilibrium points in bimatrix games. *Teorija verojatnostei i ee primemenija* 3, 1958, 318–331 (in Russian). English translation: *Theory of Probability and its Applications* 3, 1958, 297–309.

75. Vorob'ev N.N., Reduced strategies in games in extensive form, *Doklady Akademii Nauk SSSR* 115, 1957, 855–857 (in Russian).

76. Vorob'ev N.N., On decomposed strategies, *Theorija Verojatnostej i ee Primenenija* 5, 1960, 457–459 (in Russian). English translation: *Theory of Probability and its Applications* 5, 1960, 415–417.

77. Vorob'ev N.N., On coalitional games, *Doklady Akademii Nauk SSSR* 124, 1959, 253–256 (in Russian).

78. Vorob'ev N.N., Stable outcomes in coalitional games, *Doklady Akademii Nauk SSSR* 131, 1960, 493–495 (in Russian).

79. Vorob'ev N.N., The principle of optimality of Nash for general bargaining problems, *Theoretiko-igrovyje Voprosyu Prinjatija Reshenij (Game-theoretical Problems of Decision Making)*, 1978, 26–38 (in Russian).

80. Vorob'ev N.N., *Foundations of Game Theory: Non-cooperative Games*, Birkhäuser Verlag, Boston, 1994.

81. Yanovskaya E.B., Quasi-strategies in positional games, *Izvestija AN SSSR, Ser. Tekhnicheskaya kybernetika*, 1970 (N1), 14–23 (in Russian). English translation: *Engineering Cybernetics* 1, 1970, 11–19.

82. Yanovskaya E.B., Solution of infinite zero-sum games in infinite-additive strategies, *Teorija Verojatnostei i ee Primememija* 15, 1970, 162–170 (in Russian). English translation: *Theory of Probability and its Applications* 15, 1970, 153–158.

83. Yanovskaya E.B., On the existence of the value in games on the unit square with discontinuous payoff functions, *Math. Operationsforsch. Statist.* 3, 1972, 91–96 (in Russian).

Part I

Noncooperative game theory and social choice

Chapter 2

A probabilistic model of social choice

E.B. Yanovskaya

Abstract: First, an axiomatization is given of one class of deterministic social choice rules which are extensions of symmetric majority rules. Secondly, an axiomatization is given for the probabilistic social choice rule such that, for every profile P of individual preferences, the binary relation $\varphi(P)$ on the set of random alternatives is generated by the expected bilinear skew-symmetric preference intensity function on the set of alternatives. Further, it is shown that the set of maximal elements of $\varphi(P)$ coincides with the set of all optimal mixed strategies of every player in the symmetric matrix game with the preference intensity as its payoff function.

Key words: Social choice problem, majority rule, probabilistic social choice rule, matrix game

1 Introduction

A social choice problem is a problem of merging individual preferences into a unique social one. As a rule, a profile of individual preference relations is defined on a finite set of alternatives. A social preference relation is a binary relation on the same set. The rules of transformations of individual preference relations into a social one are called social choice rules. The main problems of social choice theory are 1) the proofs of existence or non-existence of social choice rules satisfying some systems of axioms characterizing properties of individual and social preferences and 2) characterizations of social choice rules if they exist. The main theorem of social choice theory –Arrow's impossibility theorem (Arrow, 1963)– asserts that if both individual and social preferences are weak orderings, then there are no social choice rules satisfying four axioms called independence, monotonicity, non-imposedness, and non-dictatorship.

The next development of the theory of social choice looked for weakenings of the conditions of Arrow's theorem providing the existence of social choice rules. For example, if arbitrary non-transitive preference relations are feasible as social preference relations, then we obtain the majority rule as a rule satisfying all other Arrow's axioms. However, non-transitive binary relations are not satisfactory as social preference relations because there may not exist the best or even maximal elements of such relations which could be considered as social choices.

One of the possible ways to settle of Arrow's paradox consists in an extension of the set of alternatives to the set of probabilistic measures (lotteries) on it. A social preference relation should be a binary relation possessing maximal elements on the set of lotteries. In both decision theory and game theory such mixed extensions of binary relations and strategy sets are well-known. In both cases the probabilistic approach to a solution concept permits to implement principles which are impossible with respect to the original model. Examples are the minimax theorem for matrix games (von Neumann and Morgenstern, 1944) and theorems about existence of maximal elements of mixed extensions of binary relations (Kiruta et al., 1980).

In social choice theory there are many papers devoted to the probabilistic approach to the social choice rules (see, e.g. Fishburn, 1972; Intrilligator, 1973; Zeckhauser, 1969). However, they are based on expected utility theory, which is applicable only for transitive preferences. If we consider non-transitive preference relations as social choice rules, e.g., majority rules, then it is necessary to apply non-linear expected utility theory for the definition of preference relations on the sets of lotteries. The first result in this direction was the paper by Kiruta (1978), who defined a mixed extension of the symmetric majority rule. The mixed extension was defined with help of the skew-symmetric bilinear theory of expected utility (Kiruta et al., 1980). This extension possessed maximal elements on any finite set of alternatives.

Next, Fishburn (1984), independently of Kiruta et al. (1980), developed the skew-symmetric bilinear theory of expected utility, based on cardinal intensities of preferences. He defined and axiomatically characterized a class of probabilistic social choice rules which were extensions of symmetric majority rules and possessed maximal elements on every finite set of alternatives.

In this paper the class of probabilistic choice rules is narrowed down to a unique probabilistic rule defined by the following cardinal estimate of the preference intensity for the symmetric majority rule: preference intensity of an alternative x with respect to another alternative y is equal to the difference of the number of individuals preferring x to y and the number of individuals preferring y to x.

2 The model

Let X be a nonempty, finite set of *alternatives* and denote by $\mathcal{P}(X)$ the set of all weak orderings on X. A binary relation R on X is called *symmetric* and *asymmetric* respectively if it holds

$$xRy \iff yRx \quad \text{for all } x, y \in X, \text{ respectively,}$$

$$xRy \implies y\bar{R}x \quad \text{for all } x, y \in X,$$

where $\bar{R}$ denotes the *negation* of the binary relation R on X. The set $c(X, R)$ of *maximal elements* of the binary relation R on X is given by $c(X, R) = \{x \in X \mid y\bar{R}x \text{ for all } y \in X, y \neq x\}$.

Let N be a finite set of n individuals ($n \geq 2$). Individual preferences are listed by $P_i \in \mathcal{P}(X)$ for all $i \in N$ and the ordered n-tuple $P = (P_i)_{i \in N} \in \mathcal{P}^n(X)$ of individual preferences is called a *profile* on X. A *social choice problem* is an ordered triple $\langle N, X, P \rangle$ where $P \in \mathcal{P}^n(X)$.

Let $\mathcal{R}(X)$ denote the set of all asymmetric binary relations on X. A *social choice rule* is a mapping $\varphi : \mathcal{P}^n(X) \to \mathcal{R}(X)$ so that the rule φ assigns to every profile $P \in \mathcal{P}^n(X)$ an asymmetric binary relation $\varphi(P)$ on the set X of alternatives. For any profile P, denote by $I(P)$ the *indifference relation* corresponding to the binary relation $\varphi(P)$ on X, i.e.,

$$xI(P)y \iff \left[x\overline{\varphi(P)}y \quad \text{and} \quad y\overline{\varphi(P)}x \right] \quad \text{for all } x, y \in X.$$

Let $\mathbb{Z}$ denote the set of all integers and $\mathbb{N}$ the set of non-negative integers (or natural numbers inclusive of zero). With every profile $P = (P_i)_{i \in N} \in \mathcal{P}^n(X)$, there are associated the following functions $n_P : X \times X \to \mathbb{N}$ and $u_P : X \times X \to \mathbb{Z}$: for all $(x, y) \in X \times X$

$$n_P(x, y) = |\{i \in N \mid xP_iy\}| \quad \text{and} \quad u_P(x, y) = n_P(x, y) - n_P(y, x).$$

In words, the preference intensity $u_P(x,y)$ of an alternative x against another alternative y equals the difference of the number of individuals preferring x to y and the number of individuals preferring y to x.
The *simple symmetric majority rule* $\varphi_{ssm} : \mathcal{P}^n(X) \to \mathcal{R}(X)$ is given, for every profile P, by

$$x\varphi_{ssm}(P)y \iff u_P(x,y) > 0 \quad \text{for all } x, y \in X.$$

Generally speaking, a *symmetric majority rule* $\varphi_{sm} : \mathcal{P}^n(X) \to \mathcal{R}(X)$ is given, for every profile P, by

$$x\varphi_{sm}(P)y \iff u_P(x,y) > \alpha \quad \text{for some } \alpha \geq 0 \quad \text{for all } x, y \in X.$$

Recall that a *majority rule* $\varphi_m : \mathcal{P}^n(X) \to \mathcal{R}(X)$ is given, for every profile P, by

$$x\varphi_m(P)y \iff n_P(x,y) > \alpha \cdot n \text{ for some } 0 < \alpha < 1, \text{ for all } x, y \in X.$$

Let $\mathcal{M}(X)$ denote the set of all *probabilistic measures* on the finite set X. A *probabilistic social choice rule* is a mapping $\varphi : \mathcal{P}^n(X) \to \mathcal{R}(\mathcal{M}(X))$ so that the rule φ assigns to every profile $P \in \mathcal{P}^n(X)$ an asymmetric binary relation $\varphi(P)$ on the set $\mathcal{M}(X)$ of so-called random alternatives.
Fishburn(1984) introduced the method of maximal lotteries in the setting of the collection $\mathcal{F}$ of all odd and monotonically increasing functions $f : \mathbb{Z} \to \mathbb{R}$ satisfying $f(1) = 1$
With every function $f \in \mathcal{F}$, there is associated a probabilistic social choice rule $\varphi_f : \mathcal{P}^n(X) \to \mathcal{R}(\mathcal{M}(X))$ as follows: for all $P \in \mathcal{P}^n(X)$ and all $\mu, \nu \in \mathcal{M}(X)$

$$\mu\varphi_f(P)\nu \iff \sum_{(x,y)\in X\times X} f(u_P(x,y)) \cdot \mu(x) \cdot \nu(y) > 0. \tag{2.1}$$

Fishburn (1984) showed that, for every $f \in \mathcal{F}$, the probabilistic social choice rule φ_f has maximal elements on every finite set X of alternatives, and an axiomatic characterization of this rule was treated, one of the axioms being as follows:

$$x\varphi_f(P)y \iff f(u_P(x,y)) > 0 \text{ for all } x, y \in X, y \neq x. \tag{2.2}$$

This axiom is not very convincing since it is nothing but the definition of the simple symmetric majority rule φ_{ssm}.

The main result of the paper is an axiomatic characterization of the probabilistic social choice rule φ_f for $f(z) \equiv z$ (Theorem 4.2). Some of the proposed axioms coincide with Fishburn's ones. However, axiom defined in (2.2) is not included, all the axioms concern with the properties of the social choice rule φ_f both on the initial set X and on the set of random alternatives $\mathcal{M}(X)$. Besides, an intermediate result – an axiomatic characterization of one class of deterministic social choice rules which are extensions of simple majority rules (Theorem 3.2) – is given.

3 Axiomatization of majority-type social choice rules

In this section we present a system of five axioms for social choice rules, which turn out to characterize a collection of social choice rules including the symmetric majority rules.

Definition 3.1 Let $\varphi : \mathcal{P}^n(X) \to \mathcal{R}(X)$ denote a generic social choice rule on X.

(i) *Binary Independence.* Let two profiles $P = (P_i)_{i\in N}$ and $P' = (P'_i)_{i\in N}$ on X satisfy
$xP_iy \Longleftrightarrow xP'_iy$ for all $i \in N$. Then $x\varphi(P)y \Longleftrightarrow x\varphi(P')y$

(ii) *Neutrality.* Let $\sigma : X \to X$ be a permutation of the set X of alternatives. Let two profiles $P = (P_i)_{i\in N}$ and $P' = (P'_i)_{i\in N}$ on X satisfy $xP_iy \Longleftrightarrow (\sigma(x))P'_i(\sigma(y))$ for all $i \in N$. Then $x\varphi(P)y \Longleftrightarrow (\sigma(x))\varphi(P')(\sigma(y))$

(iii) *Anonymity.* Let $\pi : N \to N$ be a permutation of the set N of individuals interchanging two individuals $i, j \in N$, i.e., $\pi(i) = j$, $\pi(j) = i$, and $\pi(k) = k$ for all $k \in N$, $k \neq i$, $k \neq j$. Then $\varphi(P) = \varphi(\pi P)$, where profile $\pi P = (P_{\pi k})_{k\in N}$.

(iv) *Non-imposedness.* For each pair of alternatives $x, y \in X$, there exists a profile $P \in \mathcal{P}^n(X)$ such that $x\varphi(P)y$.

(v) *Monotonicity.* Let two profiles $P = (P_i)_{i\in N}$ and $P' = (P'_i)_{i\in N}$ on X satisfy either

$$\Big[xP_iy \Longleftrightarrow xP'_iy\Big] \text{ or } \Big[yP'_ix \Longrightarrow yP_ix\Big] \quad \text{for all } i \in N.$$

Then $x\varphi(P)y \Longrightarrow x\varphi(P')y$.

These five axioms are well-known in social choice theory. Binary independence is one of the main axioms and it states that the social choice between two alternatives depends only on the individual preferences on these alternatives. Neutrality and anonymity state independence of the social choice on permutations of the sets of alternatives and individuals respectively. Non-imposedness and (a weaker form of) monotonicity are used in Arrow's theory. All five axioms imply the fulfilment of all conditions of Arrow's theorem except for transitivity of the binary relation $\varphi(P)$.

For each $n \in \mathbb{N}$ denote $A_n = \{(a,b) \in \mathbb{Z} \times \mathbb{Z} \,|\, a \geq 0, b \geq 0, a+b \leq n\}$.

Theorem 3.2 *If a social choice rule $\varphi : \mathcal{P}^n(X) \to \mathcal{R}(X)$ satisfies all five axioms, then for all profiles $P \in \mathcal{P}^n(X)$ and all $x, y \in X$*

$$x\varphi(P)y \iff g_n(n_P(x,y), n_P(y,x)) > 0 \tag{3.1}$$

for some odd function $g_n : A_n \to \mathbb{R}$ not equal to zero and such that the inequality $g_n(a,b) > 0$ implies that when the first variable increases or/and the second variable decreases, the function g_n does not change the sign.

Proof. Let $P \in \mathcal{P}^n(X)$ be any profile on X.
The three axioms binary independence, neutrality, and anonymity yield that the relation $\varphi(P)$ between any two alternatives $x, y \in X$ only depends on the numbers $n_P(x,y)$ and $n_P(y,x)$. Hence, there exists a function $g_n : A_n \to \mathbb{R}$ such that (3.1) holds. Further, the asymmetry of the relation $\varphi(P)$ implies that there exists a skew-symmetric function g_n, with respect to x and y, satisfying (3.1). Thus, such a function g_n is skew-symmetric also with respect to $a = n_P(x,y)$ and $b = n_P(y,x)$. Moreover, by the neutrality axiom,

$$\Big[n_P(x,y) = n_P(y,x)\Big] \implies \Big[x\varphi(P)y \iff y\varphi(P)x\Big]$$

From this, together with the asymmetry of the relation $\varphi(P)$, it follows

$$\Big[n_P(x,y) = n_P(y,x)\Big] \implies \Big[x\overline{\varphi(P)}y \quad \text{and} \quad y\overline{\varphi(P)}x\Big]$$

Since the function g_n is skew-symmetric, $g_n(a,a) = 0$ for all $(a,a) \in A_n$. Further, the non-imposedness axiom implies that the function g_n is not equal to the zero function. Finally, the monotonicity axiom implies the latter statement concerning the invariance of the sign of g_n whenever $g_n(a,b) > 0$. Note that the symmetric majority rules arise in case the function g_n has the form $g_n(a,b) = a - b$ whenever $|a-b| \geq \beta$ for some $0 < \beta < n$, and $g_n(a,b) = 0$ otherwise. □

4 Axiomatization of a probabilistic social choice rule

Recall that $I(P)$ denotes the indifference relation corresponding to the binary relation $\varphi(P)$ on the set $\mathcal{M}(X)$ of random alternatives. First we add three axioms for the binary relations $\varphi(P)$ or $I(P)$, where the underlying profiles $P \in \mathcal{P}^n(X)$ refer to the same set N of individuals. In the setting of variable sets of individuals, the last two axioms are concerned with properties of the sets $c(\mathcal{M}(X), \varphi(P))$ of maximal elements of any binary relation $\varphi(P)$ on $\mathcal{M}(X)$.

Definition 4.1 Let $\varphi : \mathcal{P}^n(X) \to \mathcal{R}(\mathcal{M}(X))$ denote a generic probabilistic social choice rule on X.

(vi) *Convexity and concavity.* For all $0 < \alpha < 1$, the following three implications hold:

$$\left[\mu_1 \varphi(P) \nu \text{and} \mu_2 (\varphi(P) \vee I(P)) \nu\right] \Longrightarrow \left[(\alpha\mu_1 + (1-\alpha)\mu_2) \varphi(P) \nu\right]$$

$$\left[\mu \varphi(P) \nu_1 \text{and} \mu (\varphi(P) \vee I(P)) \nu_2\right] \Longrightarrow \left[\mu \varphi(P) (\alpha\nu_1 + (1-\alpha)\nu_2)\right]$$

$$\left[\mu_1 I(P) \nu \text{and} \mu_2 I(P) \nu\right] \Longrightarrow \left[(\alpha\mu_1 + (1-\alpha)\mu_2) I(P) \nu\right]$$

(vii) *Symmetry (Fishburn, 1972).* If $\mu\varphi(P)\xi\varphi(P)\nu$ and $\xi I(P)(\frac{\mu+\nu}{2})$, then for all $0 < \alpha < 1$

$$\left[(\alpha\mu + (1-\alpha)\nu) I(P) (\tfrac{\mu+\nu}{2})\right] \Longleftrightarrow \left[(\alpha\nu + (1-\alpha)\mu) I(P) (\tfrac{\nu+\xi}{2})\right]$$

(viii) *Continuity.* If $\mu\varphi(P)\xi\varphi(P)\nu$, then there exists $0 < \alpha < 1$ such that $\xi I(P)(\alpha\mu + (1-\alpha)\nu)$
(the uniqueness of α only holds by further conditions on $\varphi(P)$).

(ix) *Independence of the number of individuals.* Let $\langle N, X, P\rangle$ and $\langle N', X, P'\rangle$ be two social choice problems with different sets of individuals. If $n_P(x,y) = n_{P'}(x,y)$ for all $x, y \in X$, then $c(\mathcal{M}(X), \varphi(P)) = c(\mathcal{M}(X), \varphi(P'))$.

(x) *Additivity.* Let $\langle N, X, P\rangle$ and $\langle N', X, P'\rangle$ be two social choice problems with different sets of individuals. Denote the union of these two problems by the social choice problem $\langle N \cup N', X, P \cup P'\rangle$ where the profile $P \cup P' = ((P_i)_{i\in N}, (P'_j)_{j\in N'})$. Then the following inclusion holds: $c(\mathcal{M}(X), \varphi(P)) \cap c(\mathcal{M}(X), \varphi(P')) \subseteq c(\mathcal{M}(X), \varphi(P \cup P'))$.

The first three axioms can be considered as an extension of von Neumann-Morgenstern's axioms of expected utility (1944) and these three axioms were treated by Fishburn (1982). The convexity/concavity axiom is well-known in the theory of binary relations in that all sections of graphs $\varphi(P)$ as subsets of $\mathcal{M}(X) \times \mathcal{M}(X)$ on each set $\mathcal{M}(X)$ are required to be convex sets. The symmetry axiom has been introduced by Fishburn (1972) and the continuity axiom is a direct analogue of the continuity axiom in the expected utility theory. The additivity axiom was introduced by P.Young (1975) under the name the reinforcement axiom. It can be considered as an extension of the unanimity axiom, which requires the inclusion $\cap_{i\in N} c(X, P_i) \subseteq c(\mathcal{M}(X), \varphi(P))$, where $P = (P_i)_{i\in N}$. In words, if some alternative is the best for each individual, then this alternative is the most preferable for the society. Note that the independence axiom, applied with equal sets of individuals, follows from the three axioms binary independence, neutrality, and anonymity.

Theorem 4.2 *If a probabilistic social choice rule* $\varphi : \mathcal{P}^n(X) \to \mathcal{R}(\mathcal{M}(X))$ *satisfies all ten axioms, then it holds* $\left[\mu\varphi(P)\nu \iff \sum_{x,y\in X} u_P(x,y)\mu(x)\nu(y) > 0\right]$ *for all profiles* $P \in \mathcal{P}^n(X)$ *and* $\mu, \nu \in \mathcal{M}(X)$ *such that the set of maximal elements* $c(\mathcal{M}(X), \varphi(P))$, *coincides with the set of optimal mixed strategies of every player in the symmetric matrix game with the payoff matrix* $\left[u_P(x,y)\right]$.

Proof. The proof of Theorem 4.2 follows Fishburn's approach (1982). Let $P \in \mathcal{P}^n(X)$ be a profile on X. If a binary relation $\varphi(P)$ on the set $\mathcal{M}(X)$ of random alternatives satisfies three axioms convexity/concavity, symmetry, and continuity, then it holds for all $\mu, \nu \in \mathcal{M}(X)$

$$\mu\varphi(P)\nu \iff W_P(\mu,\nu) := \sum_{(x,y)\in X\times X} w_p(x,y)\cdot\mu(x)\cdot\nu(y) > 0 \qquad (4.1)$$

for a certain skew-symmetric function $w_P : X \times X \to \mathbb{R}$ such that $x\varphi(P)y$ iff $w_P(x,y) > 0$ for all $x, y \in X$. The function w_P is unique up to a positive multiplier. By the definition of the set $c(\mathcal{M}(X), \varphi(P))$ of maximal elements and the skew-symmetry of the function W_P, the following equivalences hold:

$$\begin{aligned}
\mu^* \in c(\mathcal{M}(X), \varphi(P)) &\iff \nu\overline{\varphi(P)}\mu^* \quad \text{for all } \nu \in \mathcal{M}(X),\\
&\overset{(4.1)}{\iff} W_P(\nu, \mu^*) \leq 0 \quad \text{for all } \nu \in \mathcal{M}(X).\\
&\iff W_P(\mu^*, \nu) \geq 0 \quad \text{for all } \nu \in \mathcal{M}(X),
\end{aligned}$$

Hence, $\mu^* \in c(\mathcal{M}(X), \varphi(P))$ if and only if $\mu^* \in \mathcal{M}(X)$ is an optimal strategy of every player in the symmetric matrix game with payoff function $w_p : X \times X \to \mathbb{R}$.

It remains to show that the functions w_P and u_P coincide on $X \times X$. Recall, by Theorem 3.2, that the five axioms binary independence, neutrality, anonymity, non-imposedness, and monotonicity imply that (3.1) holds. Together with (4.1) (by Fishburn's approach due to the three axioms convexity/concavity, symmetry, and continuity), it follows that for any profile P

$$w_P(x, y) = g_n(n_P(x, y), n_P(y, x)) \quad \text{for all } x, y \in X$$

where the function $g_n : A_n \to \mathbb{R}$ satisfies three additional properties as described in Theorem 3.2. By the independence axiom of the number of individuals, the index n is inessential and will be omitted in the sequel. Consider the two social choice problems $\langle N, X, P\rangle$ and $\langle \{1, 2\}, X, P'\rangle$ where $P'_1 = (P'_2)^{-1}$. By (3.1), applied to $\varphi(P')$, its indifference relation $I(P')$ satisfies $xI(P')y$ for all $x, y \in X$ and therefore, $c(\mathcal{M}(X), \varphi(P')) = \mathcal{M}(X)$. Clearly, by the additivity axiom, the inclusion $c(\mathcal{M}(X), \varphi(P)) \subseteq c(\mathcal{M}(X), \varphi(P \cup P'))$ holds. Thus, for all $\mu^* \in c(\mathcal{M}(X), \varphi(P))$

$$\left[\sum_{x \in X} g(n_P(x, y), n_P(y, x)) \cdot \mu^*(x) \geq 0 \quad \text{for all } y \in X\right] \iff$$

$$\left[\sum_{x \in X} g(n_{P'}(x, y) + 1, n_{P'}(y, x) + 1) \cdot \mu^*(x) \geq 0 \quad \text{for all } y \in X\right]$$

In words, the latter equivalence means that the social choice rule φ does not depend on both numbers $n_P(x, y)$ and $n_P(y, x)$, but only on their difference $n_P(x, y) - n_P(y, x)$, that is $u_P(x, y)$. In this context, denote the function ψ by

$$\psi(\mu_P(x, y)) = g(n_P(x, y), n_P(y, x)) = w_P(x, y) \quad \text{for all } x, y \in X.$$

By skew-symmetry of the function $w_P : X \times X \to \mathbb{R}$, we obtain that for all $x, y \in X$

$$\psi(-u_P(x,y)) = \psi(u_P(y,x)) = w_P(y,x) = -w_P(x,y) = -\psi(u_P(x,y)).$$

From $\psi(u_P(x,y)) = w_P(x,y)$ and the additivity axiom, it follows that any optimal mixed strategy of both players in two symmetric matrix games with payoff functions $\psi(u_P(x,y))$ and $\psi(u_{P'}(x,y))$ respectively on $X \times X$, this mixed strategy is also optimal for both players in the symmetric matrix game with payoff function $\psi(u_P(x,y) + u_{P'}(x,y))$ on $X \times X$. In the sequel we aim to prove that this property for symmetric matrix games holds only if $\psi(z) = \alpha \cdot z$ for some $\alpha > 0$, all $z \in \mathbb{Z}$. Obviously, any function of the form $\psi(z) = \alpha \cdot z$ (for some $\alpha > 0$) fulfils the additivity axiom since, if the same mixed strategy is optimal in two symmetric matrix games of the same size, then this mixed strategy is also optimal in the sum of the two matrix games. Consider the symmetric 3×3-matrix game with the following payoff matrix

$$A_{a,b,c} = \begin{bmatrix} 0 & -\psi(c) & \psi(b) \\ \psi(c) & 0 & -\psi(a) \\ -\psi(b) & \psi(a) & 0 \end{bmatrix} \text{ where } \psi(a), \psi(b), \psi(c) > 0.$$

Obviously, the (unique) optimal mixed strategy for both players in this symmetric 3×3-matrix game $A_{a,b,c}$ is given by $\frac{(\psi(a),\psi(b),\psi(c))}{\psi(a)+\psi(b)+\psi(c)}$. Similarly, for each $n \in \mathbb{Z}$, $n \geq 1$, the (unique) optimal mixed strategy for both players in the symmetric 3×3-matrix game $A_{na,nb,nc}$ is given by $\frac{(\psi(na),\psi(nb),\psi(nc))}{\psi(na)+\psi(nb)+\psi(nc)}$. Due to the independence axiom of the number of individuals, all these symmetric 3×3-matrix games $A_{na,nb,nc}$ ($n \in \mathbb{Z}$, $n \geq 1$), have the same optimal mixed strategy. Hence, it holds

$$\frac{\psi(na)}{\psi(a)} = \frac{\psi(nb)}{\psi(b)} = \frac{\psi(nc)}{\psi(c)} \quad \text{provided } \psi(a) > 0,\ \psi(b) > 0,\ \psi(c) > 0.$$

Choosing $a = n \cdot b$ yields the equation $[\psi(n \cdot b)]^2 = \psi(n^2 \cdot b) \cdot \psi(b)$ for all $n \in \mathbb{Z}$, $n \geq 1$, provided $\psi(b) > 0$. Recall that $\psi(n \cdot b) > 0$ whenever $\psi(b) > 0$. By Aczel's (1966) theory, the functional equation $[\psi(n \cdot b)]^2 = \psi(n^2 \cdot b) \cdot \psi(b)$ for the function $\psi : \mathbb{Z} \to \mathbb{R}$ is solvable and all solutions of the equation have the form $\psi(z) = [\frac{z}{b}]^\lambda$ where $\lambda \in \mathbb{R}$ is such that $\psi(-z) = -\psi(z)$ holds. Without going into deep technical details, we claim that $\lambda = 1$ as well as $b = 1$, and so, $\psi(z) = z$ for all $z \in \mathbb{Z}$. Hence, $w_P(x,y) = \psi(u_P(x,y)) = u_P(x,y)$ for all $(x,y) \in X \times X$ and the proof is complete. □

Remark: the last part of the technical proof ($\lambda = 1$ as well as $b = 1$) is omitted here.

The proofs of these claims are fulfilled by counterexamples for the values $\lambda \neq 1$, $b \neq 1$.

Acknowledgement
This chapter is the author's translation of E.B. Yanovskaya, "Veroyatnostnaya model grouppovogo vybora," in: "Methods of data analysis, estimation and choice in system research," Annals of VNIISI, 1986, 14, 57–65 (in Russian).

References

1. Aczel, J. (1966): *Functional equations and their applications.* New York, London: Acad.Press.

2. Arrow, K.J. (1963): *Social Choice and Individual Values.* New York: Wiley (2nd ed.).

3. Fishburn, P.C. (1972): "Lotteries and social choice," *Journal of Economic Theory*, 5, 189–207.

4. Fishburn, P.C. (1982): "Nontransitive measurable utility," *Journal of Mathematical Psychology*, 26, 31–67.

5. Fishburn, P.C. (1984): "Probabilistic social choice based on simple voting comparisons," *Review of Economic Studies*, 51, 683–692.

6. Intriligator, M.D. (1973): "A probabilistic model of social choice," *Review of Economic Studies*, 40, 553–560.

7. Kiruta, A.Ya. (1978): "Axiomatic utility theory for nontransitive preferences and social choice," in: *Prikladnoi matematicheskii analiz*, Moscow, Nauka, 321–326 (in Russian).

8. Kiruta, A.Ya., A.M. Rubinov, and E.B. Yanovskaya (1980): "Optimal choice of distributions in socio-economic problems," Leningrad, *Nauka* (in Russian).

9. von Neumann, J., and O. Morgenstern (1944): *Theory of Games and Economic Behavior.* Princeton: Princeton University Press.

10. Young, P. (1975): "Social choice scoring functions. SIAM Journal of Applied Mathematics," 28, 824–838.

11. Zeckhauser, R. (1969): "Majority rule with lotteries on alternatives," *Quarterly Journal of Economics*, 83, 696–703.

Chapter 3

Equilibrium points in general noncooperative games and their mixed extensions

E.B. Yanovskaya

Abstract: General ordinal noncooperative games with a finite number of players are considered. Existence of equilibrium points in such games is proved. The definitions of different mixed extensions of these games are discussed, and the corresponding existence theorems are established. A connection between so defined mixed extensions and mixed extensions of noncooperative games with payoff functions is investigated.

Key words: Noncooperative game, preference profile, equilibrium point, mixed extension of binary relation

1 Introduction

Non-cooperative games are usually defined in the payoff function form, i.e. real-valued payoff functions are supposed to be defined for every player. In general noncooperative games only preference relations of players are given. The first definition of such games belongs to Farquharson (1955).

The main tool for the proof of existence of equilibrium points in noncooperative games is fixed point theorems. The applicability of such theorems demands, in particular, quasi-concavity of players' payoff functions in the strategies of the corresponding player. This property is fulfilled in the mixed extensions where payoff functions, being mathematical expectations of payoffs, are linear in the mixed strategies, i.e. probability distributions.

In order to define mixed extensions of general noncooperative games it is necessary to define preference relations of players on the outcomes, i.e. on the product of mixed strategies of all players.

This paper is a review of the author's papers (1973, 1974, 1976), devoted to the solution of this problem for different preference profiles.

2 Existence of equilibrium points in general noncooperative games

A general noncooperative game in the normal form is a triple

$$\Gamma = \langle N, \{X_i\}_{i\in N}, \{R_i\}_{i\in N}\rangle, \tag{2.1}$$

where N is a finite set of players, X_i is a set of strategies of the player i, R_i a binary (preference) relation of the player i, defined on the product $X = \prod_{i\in N} X_i$. We suppose in the sequel that the relations R_i are reflexive, i.e. they are non strict preference relations. The n-tuple $(R_1, \dots, R_n)$ is called the *preference profile.*

Any relation R can be presented as $P \vee I$, where P and I are its asymmetric and symmetric parts respectively. In the sequel we denote $P = P(R)$, $I = I(R)$, if it is necessary to show the link between the relations. Besides, for an arbitrary binary relation R, we denote by $J(R)$ the incomparability relation, generated by R :

$$xJ(R)y \iff [x\bar{R}y \ \text{ and } \ y\bar{R}x],$$

where the line over a relation sign denotes the negation of the relation.

A binary relation R on an arbitrary set is called a *preference ordering (ordering)*, if it is complete, reflexive, and transitive. The relation $P(R)$ is called the *strict ordering*, and $I(P)$ is an equivalence relation.

If the relations R_i for all $i \in N$ represent the order of the real line, i.e. there exist functions H_i, defined on the set X, such that

$$xR_iy \iff H_i(x) \geq H_i(y),$$

then the game Γ in (2.1) is a *game with payoff functions* and denoted by

$$\Gamma = \langle N, \{X_i\}_{i\in N}, \{H_i\}_{i\in N}\rangle. \tag{2.2}$$

The function H_i is called the *payoff function of the player i.*

The main solution concept for the games (2.2) is Nash equilibrium (Nash, 1951). An n-tuple $x^* \in X$ is called an *equilibrium point*, if for all $x_i \in X_i$ and $i \in N$

$$H_i(x^*) \geq H_i(x^* \| x_i),$$

where $x^* \| x_i = (x_1^*, \ldots, x_{i-1}^*, x_i, x_{i+1}^*, \ldots x_n^*)$.

This definition is directly extended to the general noncooperative games as follows: an n-tuple is called an equilibrium point in the game (2.1), if for all $x_i \in X_i$ and $i \in N$ it holds

$$(x^* \| x_i) \bar{P}_i x^*. \tag{2.3}$$

As it has been already noted, the key tool for proving existence of equilibrium points in games (2.2) is fixed point theorems. The similar method can be applied for general noncooperative games as well. For this purpose it is necessary to provide the strategy sets with some topological and algebraic conditions.

Let S be a linear space. A binary relation R on S is called *concave (convex)* if

$$x_1 R y,\ x_2 R y \implies (\alpha x_1 + (1-\alpha) y_2) R y \text{ for all } \alpha \in (0,1), \tag{2.4}$$

$$(x R y_1,\ x R y_2 \implies x R(\alpha y_1 + (1-\alpha) y_2) \text{ for all } \alpha \in (0,1)). \tag{2.5}$$

If the relation R is transitive, then the definitions (2.4), (2.5) are equivalent to the following ones:
a transitive relation R is called *concave (convex)*, if

$$xRy \iff (\alpha x + (1-\alpha) y) R y,\ (xRy \implies xR(\alpha x + (1-\alpha) y))$$

for all $\alpha \in (0,1)$.

Let now S be a topological space. If R is a weak or linear ordering on S, then R is called *upper semicontinuous*, if the sets

$$\{x \in S \mid xRy\} \text{ are closed for all } y \in S, \tag{2.6}$$

or

$$\{x \in S \mid yPx\} \text{ are open for all } y \in S. \tag{2.7}$$

Similarly, a relation R is called *lower semicontinuous*, if the sets

$$\{x \mid yRx\} \text{ are closed for all } y \in S, \tag{2.8}$$

or

$$\{x \in S \mid xPy\} \text{ are open for all } y \in S. \tag{2.9}$$

A preference ordering R is called *continuous*, if it is both upper and lower semicontinuous. It is easy to note that if the topology in S is generated by the intervals $(a, b) = \{x \in S \mid bPxPa\}$, then the definitions of semicontinuity and continuity of a binary relation are equivalent to that of semicontinuity and continuity of a set-valued mapping $R : S \to S$, where $y \in R(x)$, if yRx.

In the general case, when R is not an ordering, the relations (2.6),(2.7) and (2.8),(2.9) are not equivalent and the fulfilments of (2.6) and (2.8) or (2.7) and (2.9) are not the definitions of semicontinuity and continuity of a binary relation respectively.

We recall one more property of binary relations defined on topological spaces. A binary relation on S *is preserving in the limit*, if each its linear ordered component is a closed subset of S. This definition was introduced by Gillies (1959) for Euclidean spaces and was generalized to arbitrary topological spaces by Kulakovskaya (1976). Gillies proved that if a relation is preserving in the limit, then it has maximal and minimal elements on each compact set.

Theorem 2.1 *If a set of players N is finite, the strategy sets X_i are compact subsets of locally convex topological spaces, the relations $P_i = P_i(R_i)$ are acyclic, the relations $\bar{P}_i$ are convex and the sets*

$$\{(x, x^i) \mid xP_i(x\|y_i)\} \tag{2.10}$$

are open in $X \times X^i$, where $X^i = \prod_{j \neq i} X_j, x^i \in X^i$, then in the game Γ (2.1) there are equilibrium points.

Proof. This is proved with the help of a fixed point theorem. Let $T : X \to X$ be the best-response mapping, defined by $y \in Tx$, if

$$y_i \in \{\xi_i \in X_i \mid (x\|z_i)\bar{P}_i(x\|\xi_i) \text{ for all } z_i \in X_i\}.$$

The opennes of the sets (2.10) and acyclicity of the relations P_i imply that the relations $\bar{P}_i$ are preserving in the limit. Therefore, the sets Tx are not empty for all $x \in X$.

The convexity of the relations $\bar{P}_i$ implies that of the sets Tx. By the condition (2.10) the mapping T is closed, hence, Glicksberg's theorem (1952) implies existence of a fixed point $x^* \in Tx^*$. It is easy to notice that X^* is an equilibrium point of Γ. □

In Theorem 2.1 the conditions are imposed both on the relations P_i and on $\bar{P}_i$. However, the conditions would be more clear if they are put only on the relation $R_i(P_i)$. It is possible when the relations R_i are complete, i.e. for all $x, y \in X$ it holds xR_iy or yR_ix. In fact, in this case convexity of $\bar{P}_i$ is equivalent to concavity of R_i. However a complete reflexive acyclic relation R_i is a weak ordering. Thus, for such relations Theorem 2.1 can be formulated as follows:

Corollary 2.2 *If in a game Γ (2.1) the set of players N is finite, the strategy sets X_i are compact subsets of linear topological spaces, the preference relations R_i are orderings on X, and the sets*

$$\{(x, x^i) \mid xR_i(x\|y_i)\} \tag{2.11}$$

are closed in $X \times X^i$ for any $y_i \in X_i$, then in the game Γ there exist equilibrium points.

Note that the compactness of X_i (and, hence, of X) and the condition (2.11) imply separability of X, and, thus, on X there exist utility functions representing the relations P_i, concave in x_i and upper semicontinuous in x. The last claim is an extension of Theorem 3.5 in Fishburn (1970).

Thus, the statement of the Corollary is equivalent to the following one:

Theorem (Nikaido and Isoda, 1955) *If in a noncooperative game Γ (2.2) the set of players N is finite, the strategy sets are compact subsets of linear topological spaces, the payoff functions H_i are quasiconcave in x_i, upper semicontinuous in x and equicontinuous in x^i, then in the game Γ there are equilibrium points.*

If the preference relations R_i are complete, then definition (2.3) is equivalent to the following one: a strategy n-tuple in the game Γ (2.1) is called *an equilibrium point* if for all $i \in N$, $x_i \in X_i$

$$x^* R_i(x^* \| x_i). \tag{2.12}$$

It is clear that that if R_i are not complete than the equilibrium points in definition (2.12) are more strong concept than definition (2.1).

It is possible to introduce one more equilibrium concept connecting the two given above – (2.1) and (2.12) – which reminds of the von Neumann–Morgenstern solution for cooperative games.

Define the following binary relations $\tilde{R}_i$ on the strategy set X :

$$x\tilde{R}_iy, \quad , \text{ if } xR_i(x\|y_i),$$
$$R^1 = \bigcup_{i\in N} \tilde{R}_i, \quad R^2 = \bigcap_{i\in N} \tilde{R}_i.$$

Evidently, $xR^2y \implies xR^1y$ and the set of equilibrium points in definition (2.1) is the set $C(X, R^1)$ of maximal elements on the set X with respect to the relation R^1. The set $C(X, R^1)$ is called *externally stable* if it is stable w.r.t. R^2, i.e. for each $y \in X$ there is $x^* \in C(X, R^1)$ such that for all $i \in N$ it holds $x^* R_i(x^* \| y_i)$. The set of external equilibrium points reminds the core coinciding with the N–M solution for cooperative games, though here the set of maximums (an analogue of internal stability) is given w.r.t. the relation R^1, and the external stability w.r.t. R^2. It is easy to note that if in a game there exists an equilibrium point (2.12), then the set of its equilibrium points (2.1) is externally stable.

Theorem 2.3 *If in the conditions of Theorem 2.1 the relations R_i are concave, are preserving in the limit and the sets*

$$\{x \in X \mid xR_i(x\|y_i)\}, \quad y_i \in X_i,$$

are closed then the set of equilibrium points (2.1) of the game Γ is externally stable.

Proof. For each $y \in X$ consider the mapping $T_y : X \to X$, defined as follows:

$$(T_yx)_i = \left\{ \xi_i \in X_i \;\middle|\; \begin{array}{ll} (x\|\xi_i)R_i(x\|y_i) & \text{and } (x\|y_i)\bar{P}_i(x\|\xi_i) \\ & \text{for all } x \in X \end{array} \right\}.$$

The nonemptiness of the sets T_yx for all $x, y \in X$ follows from the preserving in the limit of the relations R_i. Further, by the same way as in the proof of Theorem 2.1 we obtain that for any $y \in X$ the mapping T_y has a fixed point $x^* \in T_yx^*$ such that x^*R^2y. □

Example. Consider a two-person zero-sum game on the unit square, having 6 different outcomes corresponding to the strategy pairs as follows:

$$(x, y) \text{ corresponds to } \begin{cases} e, & \text{if } x > y, x \neq 1, \\ a, & \text{if } x < y, y \neq 1, \\ c, & \text{if } x = y, x \neq 1, \\ b, & \text{if } x = 1, y \neq 1, \\ d, & \text{if } x \neq 1, y = 1, \\ f, & \text{if } x = 1, y = 1, \end{cases}.$$

Let the preference relation R and the strict preference $P = P(R)$ of the player 1 be the following: $eRfRa$, $ePdPcPbPa$, and in the each chain the relations are transitive.

It is clear that $(1,1)$ is the unique equilibrium point in the game. However, for the strategy pairs $\theta = (x \neq 1, 1), \zeta = (1, y \neq 1)$ the relations $\xi R^2 \theta, \xi R^2 \zeta$ do not hold, and ξ is not externally stable. Note that the relation R is not preserving in the limit. In fact, fix $y \neq 1$ and let $x_n \to 1$. Then for sufficiently large m, n we have $(x_n, y) R (x_m, y)$. However the relation $(1, y) R (x_m, y)$ is not valid.

3 Mixed extensions of noncooperative games

The proofs of existence theorems given in Section 2 demand strong enough conditions on the sets of strategies and preference profiles. In particular, in all the theorems the linearity of strategy spaces was necessary. Moreover, the condition of concavity of preference relations is enough strong.

To prove existence theorems for more large classes of games mixed extensions of original games are usually considered, where the strategy sets of initial (pure) strategies are extended to the sets of probability measures on them. Then existence theorems are proved for mixed extensions.

Thus, a mixed strategy of a player means the random choice of his pure strategy w.r.t. a probability measure, defining the mixed strategy. As the set of probability measures defined on some set is convex, then the sets of mixed strategies of the players are convex subsets of linear spaces and the corresponding conditions in the existence theorems turn out fulfilled.

To define a mixed extension of a game completely it is necessary more to define preference relations of the players on n-tuples of mixed strategies such that they would coincide with the preference relations of the original game on the n-tuples of pure strategies considered as the degenerate measures. For games (2.2) the mixed extensions of preference relations of the players are defined by theirs payoffs in mixed strategies. The last ones are supposed to be mathematical expectations of the random payoffs. Formally, let $\mu = (\mu_1, \ldots, \mu_n)-$ be a mixed strategy $n-$tuple. The payoff of the player i for this strategy $n-$tuple is

$$\begin{aligned} H_i(\mu) &= \int_X H_i(x)\, d\mu(x) \\ &= \int_{X_n} \cdots \int_{X_1} H_i(x_1, \ldots, x_n)\, d\mu_1(x_1) \ldots d\mu_n(x_n) \end{aligned} \tag{3.1}$$

if the corresponding assumptions about measurability of payoff functions are fulfilled and there exists n-times integral in (3.1). Payoffs (3.1) define the mixed extension completely.

Such a definition of a mixed extension of the games (2.2) is the most popular though not unique. For example, in the median theory of Walsch (1969) the players' preference relations in a mixed extension are defined by the numerical ordering between the medians of probability measures corresponding to the mixed strategy $n-$tuples.

Consider a general noncooperative game Γ as in (2.1).
Let $\mathcal{A}_i$ be a $\sigma-$algebra of subsets of X_i, containing all singletons and the whole set X_i. By M_i we denote the set of all probabilistic measures, defined on $\mathcal{A}_i$. A *mixed extension* of the game Γ is a game

$$\bar{\Gamma} = \langle N, \{M_i\}_{i\in N}, \{R_i\}_{i\in N}, \rangle, \tag{3.2}$$

where R_i are arbitrary reflexive binary relations on the mixed strategy set $M = \prod_{i\in N} M_i$ such that on the set $X \subset M$ the relations coincide with the ones defining the game Γ. We denote them by the same letters R_i, it should not lead to a confusion.

A mixed extension is called *complete*, if for all players $i \in N$ the relations R_i are complete on M. Evidently, the infinite number of mixed extensions correspond to each game Γ.

The *trivial* (non complete) mixed extension of Γ is the game Γ itself. In such an extension preference relations are defined only on the subset $X \subset M$. Therefore, any mixed strategy $n-$tuple from the set $\prod_{i\in N}(M_i \setminus X_i)$ is an equilibrium point in the trivial mixed extension.

An example of a complete mixed extension is the above defined mixed extension of the games with payoff functions (2.2).

Thus, the problem of the definition of a mixed extension of a noncooperative game is reduced to the one of the definition of a mixed extension of a binary relation. Some approaches to the solution of the last problem will be considered in the next section.

4 Mixed extensions of binary relations

Let X be an arbitrary set, $\succ$ be a binary relation on it. Denote by M the set of all probability measures on a $\sigma-$algebra subsets on X, containing all the singletons of X and the whole set X. Let be necessary to extend the relation $\succ$ on the set M in accordance with some axioms and such that the new extended relation $\tilde{\succ}$ on M would coincide with the $\succ$ on $X \subset N$, i.e.

$$x \tilde{\succ} y \Longleftrightarrow x \succ y \text{ for } x, y \in X.$$

An extension $\tilde{\succ}$ on M of the relation $\succ$ will be also denoted by $\succ$ and we call it a *mixed extension* of the binary relation $\succ$ on M.

We will define a mixed extension axiomatically with help of a system of axioms which, first, formalize enough natural properties of the extension, second, define the mixed extension uniquely and, third, the corresponding mixed extension provides the existence of equilibrium points for enough large classes of noncooperative games. For the last purpose the existence of maximal elements of the mixed extensions of binary relations has a big importance.

The cases of different sets X and relations $\succ$ on it we consider separately.

Case 1. X is a finite set, $\succ$ is an ordering. We will denote further such an ordering by R. Let M be the set of all probabilistic measures on X. Let us introduce an axiom, formalizing the following property of an extension of R to the set M :

Axiom 1. Linearity. If $x, y \in M$, then for all $\mu, \nu, \lambda \in M$, $\alpha \in (0, 1)$ it holds:

$$\mu P \nu \implies (\alpha\mu + (1-\alpha)\lambda) P (\alpha\nu + (1-\alpha)\lambda),$$
$$\mu I \nu \implies (\alpha\mu + (1-\alpha)\lambda I (\alpha\nu + (10\alpha)\lambda)$$

where $P = P(R), I = I(R)$. Since $X \subset M$, then Axiom 1 does provide an extension of P, I or R on some subset of the set $M \setminus X$.

If P is a strict ordering, then Fishburn (1972) calls axiom 1 the *weak individual axiom.*

Extend the relation P on the set M by axiom 1 and transitivity (this is possible because of the transitivity P on X.) The extended relation $\tilde{P}$ will be called the *linear transitive extension* of the relation P.

Evidently, for the linear transitive extension the following relation holds:

$$\mu_1 \tilde{P} \nu_1,\ \mu_2 \tilde{P} \nu_2 \implies (\alpha\mu_1 + (1-\alpha)\mu_2) \tilde{P} (\alpha\nu_1 + (1-\alpha)\nu_2), \qquad (4.1)$$

for all $\alpha \in (0, 1)$, $\mu_1, \mu_2, \nu_1, \nu_2 \in M$.

Denote by $F_\mu(z)$ the distribution function, corresponding to the measure μ :

$$F_\mu(z) = \mu\{x \in X \mid zPx \text{ or } zIx\}.$$

In the following theorem simple necessary and sufficient conditions of existence the relation $\tilde{P}$ between two measures μ, ν are ascertained:

Theorem 4.1 (Yanovskaya, 1975) $\mu \tilde{P} \nu \iff F_\mu(z) \leq F_\nu(z)$ *for all* $x \in X$, *where*
$F_\mu(z) \leq F_\nu(z) \forall z \in X \iff [F_\mu(z) \leqq F_\nu(z)$ *and* $F_\mu(z) \not\equiv F_\nu(z)]$.

Corollary 4.2 *The relation* $\tilde{P}$ *is a strict partial ordering on* M.

Corollary 4.3 *The relation* $(\alpha\mu_1 + (1-\alpha)\nu)\tilde{P}(\alpha\mu_2 + (1-\alpha)\nu)$ *for some* $\mu_1, \mu_2, \nu \in M$, $\alpha \in (0,1)$ *implies* $\mu_1 \tilde{P} \mu_2$.

Corollary 4.4 *If* $\mu\tilde{P}\nu$, $y(P\vee =)z$, $\mu, \nu \in M, z \in X$ *then* $(\alpha\mu + (1-\alpha)z)$ $\tilde{P}(\alpha\nu + (1-\alpha)y)$.

The Corollaries are proved by the direct application of Theorem 4.1.

Corollary 4.3 and the linearity axiom are equivalent to the fulfilment the *moderate individual axiom* (Fishburn, 1972) for the relation $\tilde{P}$. Thus, if the set M is a finite-dimensional simplex and the relation $\tilde{P}$ on the vertices of the simplex is a strict complete ordering, then the weak individual axiom is equivalent to the moderate one:

$$\mu\tilde{P}\nu \iff (\alpha\mu + (1-\alpha)z)\tilde{P}(\alpha\nu + (1-\alpha)z),$$

$$\mu\tilde{I}\nu \iff (\alpha\mu + (1-\alpha)z)\tilde{I}(\alpha\nu + (1-\alpha)z),$$

where $\mu, \nu \in M$ and $\tilde{I}$ is the following equivalence relation corresponding to the relation $\tilde{P}$:

$$\mu\tilde{I}\nu \iff [\mu\tilde{P}z \iff \nu\tilde{P}z \text{ for all } z \in M].$$

Case 2. X is an infinite set with a strict ordering P given on it. In the sequel we shall suppose that X is a compact topological space, M is the set of probabilistic measures, defined on the $\sigma-$algebra of the Borel subsets of X. Define on the set $K \subset M$ of discrete measures (with finite supports) the strict partial ordering $\tilde{P}$ as in the case 1. Let $\mu_n, \nu_n \in M$ be such that $\mu_n\tilde{P}\nu_n$. Let moreover μ_σ, ν_σ be converging extended subsequences of the sequences μ_n, ν_n (such extended subsequences do exist because of compactness of the set M in the weak* topology), and let $\mu = \lim \mu_\sigma, \nu = \lim \nu_\sigma$. Extend next the relation $\tilde{P}$ on the set M putting $\mu\tilde{P}\nu$ for the μ, ν defined above, $\mu \neq \nu$ (we keep for the extension the notation $\tilde{P}$).

Theorem 4.5 *If the relation* P *is upper semicontinuous, i.e. the sets* $\{x \in X \mid xPy\}$ *are open for all* $y \in X$, *then*

$$\mu\tilde{P}\nu \iff F_\mu(z) \leq F_\nu(z) \textit{ for all } z \in X.$$

Proof. For $\mu, \nu \in K$ the proof coincides with that of Theorem 2.1. Let $\mu\tilde{P}\nu$, $\mu_\sigma, \nu_\sigma \in K$ be extended sequences defining the relation $\mu\tilde{P}\nu$, i.e. $\mu_\sigma\tilde{P}\nu_\sigma$ and $\lim \mu_\sigma = \mu$, $\lim \nu_\sigma = \nu$. Since $\mu_\sigma, \nu_\sigma \in K$ we have the inequalities

$$F_{\mu_\sigma}(z) \leq F_{\nu_\sigma}(z) \text{ for all } z \in X. \tag{4.2}$$

By the conditions of the Theorem the sets $\{x \in X \,|\, yPx\}$ are open, hence, the distribution functions F_μ are upper semicontinuous (continuous from the left) for all $\mu \in M$ on X. Therefore, the inequalities (4.2) imply

$$F_\mu(z) \leq F_\nu(z) \tag{4.3}$$

for all points of continuity z of the functions $F_\mu(z)$ and $F_\nu(z)$. Therefore because of the continuity from the left of the distribution functions the inequality (4.3) holds for all $z \in X$. If $\mu = \nu$, then

$$F_\mu(z) \equiv F_\nu(z) \text{ for all } z \in X.$$

By the definition of the relation $\tilde{P}$ from $\mu\tilde{P}\nu$ it follows that $\mu \neq \nu$. Therefore, there exists $z \in X$ such that $F_\mu(z) < F_\nu(z)$.

Let now $\mu, \nu \in M$ be such that $F_\mu(z) \lneqq F_\nu(z)$ for all $z \in X$ and there is $z_0 \in X$ such that $F_\mu(z_0) < F_\nu(z_0)$. Because of continuity from the left and monotonicity of the distribution functions there is a left semi-neighborhood $U(z_0)$ of the point z_0 $(z \in U(z_0) \implies z_0Pz)$ such that $z \in U(z_0)$ implies

$$F_\mu(z) < F_\nu(z). \tag{4.4}$$

(Note that z_0 is not a minimal element of X w.r.t. P, because for any minimal element z and each measure $\mu \in M$ it holds $F_\mu(z) = 0$.)

Distribution functions can be also uniformly approximated by finite-valued distribution functions (step-functions) corresponding to discrete measures μ_n, ν_n such that for sufficiently large n $F_{\mu_n} \lneqq F_{\nu_n}(z)$ for all $z \in X$, and (4.4) implies that for sufficiently small neighborhood $U(z_0)$ $F_{\mu_n}(z) < F_{\nu_n}(z)$ for all $z \in U(z_0)$, i.e. we obtain that $\mu_n\tilde{P}\nu_n$, $\mu_n, \nu_n \in K$ and $\mu = \lim \mu_n, \nu = \lim \nu_n$, $\mu \neq \nu$ and, hence, $\mu\tilde{P}\nu$. □

Theorem 4.5 implies that the relation $\tilde{P}$ is a strict partial ordering on the set M.

Since the set X is compact in the topology in which the relation P is upper semicontinuous, the set of maximal elements of X w.r.t. the relation P (and, hence, w.r.t. $\tilde{P}$) is not empty.

Let $L \subset M$ be a closed subset being a strictly and completely ordered component of the relation $\tilde{P}$, i.e. the relation $\tilde{P}$ on the set L is a strict ordering.

Lemma 4.6 *On the set L the relation $\tilde{P}$ is upper semicontinuous in the weak* topology of the space of the regular countably additive functions rca(X).*

Proof. We must prove that the sets $\{\mu \in L \,|\, \nu\tilde{P}\mu\}$ are open for each $\nu \in L$ or, that is the same, that the sets $\{\mu \in L \,|\, \mu\tilde{R}\nu\}$ are closed, where $\tilde{R} = \tilde{P} \vee = .$

Let $\mu_\sigma \to \mu, \mu_\sigma \in L$ and $\mu_\sigma\tilde{R}\nu$. If for infinite number of σ $\mu_\sigma = \nu$, then $\mu = \nu$ and $\mu\tilde{R}\nu$. Otherwise we have $\mu_\sigma\tilde{P}\nu$ and Theorem 4.1 implies that $\mu\tilde{R}\nu$. □

In other words the statement of the Lemma means that the relation $\tilde{R}$ on the set M is preserving in the limit.

Consider a closed subset $M_1 \subset M$. Lemma 4.6 implies that the set of maximal elements $C(M_1, \tilde{P}) \neq \emptyset$. However even for convex sets M_1 the set $C(M_1, \tilde{P})$ may not be convex.

Case 3. X is an arbitrary set, $\succ$ is an arbitrary asymmetric relation on X.

For such relations the mixed extensions considered above are not in general well-defined, because the set $\tilde{P}$ on the set M may not coincide with P on the set X.

Definition 4.7 A function $K : X \times X \to I\!R^1$ *is preserving the relation* $\succ$defined on X, if

$$x \succ y \iff K(x,y) > 0 \text{ for all } x, y \in X. \tag{4.5}$$

Since the relation $\succ$ is supposed to be asymmetric, there are skew-symmetric functions, preserving it.

Let $T : X \to X$ be an arbitrary single-valued mapping.

Definition 4.8 A function $K : X \times X \to I\!R^1$ is called *homogenous in* $\succ$ if the following relations hold:

$$\{x \succ y \iff Tx \succ Ty \text{ for all } x, y \in X\} \implies K(x,y) = K(Tx, Ty)$$

$$\{x \succ y \iff Tx \prec Ty \text{ for all } x, y \in X\} \implies K(x,y) = K(Ty, Tx).$$

As M we shall consider the set of all probability measures defined on a $\sigma-$algebra $\mathfrak{X}$ of subsets of X, such that for all $y \in X$

$$\{x \in X \,|\, x \succ y\}, \;\; \{x \in X \,|\, y \succ x\} \in \mathfrak{X}$$

and $\mathfrak{X}$ contains all singletons of X and the whole set X.

We extend the relation $\succ$ on M according with the following axioms:

Axiom 2. Concavity–Convexity. For all $\mu_1, \mu_2 \in M$

$$\mu_1 \succ \nu,\ \mu_2 \succ \nu \implies (\alpha\mu_1 + (1-\alpha)\mu_2) \succ \nu \tag{4.6}$$

$$\nu \succ \mu_1,\ \nu \succ \mu_2 \implies \nu \succ (\alpha\mu_1 + (1-\alpha)\mu_2) \tag{4.7}$$

for all $\alpha \in (0,1)$.

Relation (4.6) is the definition of the concavity of the relation $\succ$, and relation (21) is the definition of its convexity. This axiom is natural enough. A less natural axiom would be the linearity axiom, considered in the Case 1. In fact, the linearity axiom is feasible for transitive relations, however in the case when we may have a cycle $x \succ y \succ z \succ x$ the implication of the relation

$$(\alpha x + (1-\alpha)z) \succ (\alpha y + (1-\alpha)z) \text{ for all } \alpha \in (0,1)$$

would be doubtful.

Axiom 3. Independence of inessential alternatives. Let $X, \succ$ be defined above, $X' \subset X$. Denote by $\succ'$ the restriction of the relation $\succ$ on X', and by M' the set of probability measures on X'. Let $\mu, \nu \in M'$, (hence, $\mu.\nu \in M$.) Then

$$\mu \succ' \nu \iff \mu \succ \nu.$$

Since a relation is completely defined by the preserving it function K, the properties of $\succ$ can be formulated as the ones of the function K. Evidently, an extension of $\succ$ on the set M, satisfying axioms 2 and 3 is not unique.

Theorem 4.9 (Yanovskaya, 1976) *Among the extensions on the set M of a relation $\succ$ defined on X, and satisfying axioms 2 and 3 there is the unique extension such that its preserving function is bilinear, skew-symmetric and homogenous.*

The proof of Theorem 4.9 implies that for this unique extension stated in the Theorem satisfies the following relation:

$$\begin{aligned} \mu \succ \nu \quad \iff \quad & \mu\nu\{(x,y) \in X \times X \mid x \succ y\} \\ & -\mu\nu\{(x,y) \mid (x,y) \in X \times X \mid y \succ x\} > 0. \end{aligned} \tag{4.8}$$

We are interested now in the non-emptiness of the set $C(M, \succ)$ for the relation $\succ$ satisfying (4.8). From the definition of the function K (4.5) and relation (4.8) it follows that

$$\mu^* \in C(M, \succ) \iff K(\mu, \mu^*) \le 0 \ \ (K(\mu^*, \mu) \ge 0) \text{ for all } \mu \in M,$$

i.e. μ^* is an optimal mixed strategy of both players in the symmetric zero-sum two-person game $\Gamma_K = \langle X, X, K \rangle$. Therefore, the problem of non-emptiness of the set $C(M, \succ)$ is reduced to the problem of the existence of optimal strategies in the game Γ_K. Unfortunately, in the general case the function K does not possess conditions sufficient for the application of the known minimax theorems. However, if the set X is finite, then the game Γ_K is a matrix game, and both players have optimal strategies in it.

With point of view of existence of maximal elements on the set M it is more convenient to consider another mixed extension $\succ'$ defined as follows:

$$\mu \succ' \nu \iff \mu\nu\{(x,y) \in X \times X \,|\, x \succ y\} - \mu\nu\{(x,y) \in X \times X \,|\, x \bar{\succ} y\} > 0.$$

Similarly to the function K we define another function H on X preserving the relation $\succ$:

$$H(x,y) = \begin{cases} 1, & \text{if } x \succ y, \\ -1 & \text{otherwise} \end{cases}.$$

Then if we suppose that the sets $\{y \in X \,|\, x \succ y\}$ are open for all $x \in X$ in the Borel topology of the space X, then the function $H(x,y)$ is lower semicontinuos in $y \in X$, and if more the set X is compact then there exists a mixed optimal strategy of the player 2 in the game $\Gamma_H = \langle X, X, H \rangle$, and its value $\mathrm{val}\Gamma_H \leq 0$.

Therefore, if ν^* is an optimal strategy of the player 2 in the game Γ_H, then $\nu^* \in C(M. \succ')$, because for all $\mu \in M$

$$H(\mu, \nu^*) = \mu\nu^*\{(x,y) \,|\, x \succ y\} - \mu\nu^*\{(x,y) \,|\, x \bar{\succ} y\} \leq \mathrm{val}\,\Gamma_H \leq 0.$$

In the next Section the defined above mixed extensions will be used to the definition of mixed extensions of noncooperative games and to the proof of the existence of equilibrium points in these mixed extensions.

5 Equilibrium points in mixed extensions of non-cooperative games

5.1 Games whose preference profiles are weak orderings

Consider a general noncooperative game Γ (2.1), where the relations R_i are weak orderings on the set $X = \prod_{i \in N} X_i$ for all $i \in N$.

As in the previous section we shall consider the cases of finite and infinite sets X_i separately.

Case 1. The sets X_i are finite for all $i \in N$. Let $R_i = P_i \vee I_i$, and let $\tilde{P}_i$ be the linear transitive extensions of P_i on the set M. Denote $\tilde{R}_i = \tilde{P}_i \vee \tilde{I}_i$, where $\tilde{I}_i$ are the following equivalence relations:

$$x\tilde{I}_i y \iff \{x\bar{\tilde{P}}_i a \iff y\bar{\tilde{P}}_i a\} \text{ for all } x, y, a \in M.$$

Then $\tilde{R}_i$ are reflexive.

Definition 5.1 The mixed extension $\bar{\Gamma}_A$ of the game Γ

$$\bar{\Gamma}_A = \langle N, \{M_i\}_{i\in N}, \{\tilde{R}_i\}_{i\in N}\rangle$$

is called the *minimal non-Archimedian extension* of the game Γ.

This definition is provided by the fact that for all $i \in N$ the relations $\tilde{R}_i$ satisfy all the axioms of the expected utility theory (von Neumann and Morgenstern, 1944) except for possibly the continuity (Archimedian) axiom. More exactly, the relations $\tilde{R}_i$ should be considered on the convex hull co$X$ of the set X. However, $\tilde{R}_i$ are quasi-orderings on the set $M \subset$ co$X$, and hence on co$X$. The extension is called "minimal" because any other mixed extension of the Γ, whose preference relations R'_i are extensions of the relations $\tilde{R}_i$:

$$x\tilde{R}_i y \implies xR'_i y, \;\; x, y \in M,$$

would demand some other conditions defining R'_i besides the definition of the game Γ itself.

Theorem 5.2 (Yanovskaya, 1974) *In the game $\bar{\Gamma}_A$ there are equilibrium points.*

Case 2. The sets X_i are infinite.(We shall not consider separately the case when some of X_i are finite, and some of them are infinite). As usual we suppose that X_i are topological spaces, M_i are the sets of all probability measures defined on $\sigma-$ algebras of Borel subsets of X_i. By $X_i|_\sim$ we denote the space X_i factorized by the equivalence relations I_i. (Recall that if R_i are orderings, then the corresponding relations I_i are equivalences). Then the game Γ can be defined equivalently (up to the existence and the equivalence of equilibrium points) as follows:

$$\Gamma' = \langle N, \{X_i|_\sim\}_{i\in N}, \{P_i\}_{i\in N}\rangle, \tag{5.1}$$

where P_i are strict orderings on $X|_\sim = \prod_{i\in N} X_i|_\sim$. If the spaces $X_i|_\sim$ are endowed by a topology in which the relations P_i are upper semicontinuous, then on the set M a strict partial ordering $\tilde{P}_i$ is defined (Theorem 4.9) for all $i \in N$ which is an extension of P_i. Consider the following mixed extension of the game Γ' :

$$\bar{\Gamma}|_\sim = \langle N, \{M_i|_\sim\}_{i\in N}, \{\tilde{P}_i\}_{i\in N}\rangle,$$

where $M_i|_\sim$ is the set of probability measures on $X_i|_\sim$.

Since we should define the games for non strict preference profile, then we obtain the following game:

$$\bar{\Gamma} = \langle N, \{M_i\}_{i\in N}, \{\tilde{R}_i\}_{i\in N}\rangle, \tag{5.2}$$

where M_i are the sets of probability measures on X_i, $\tilde{R}_i$ are the partial orderings corresponding to the strict partial ordering $\tilde{P}_i$.

It is clear that if the space $X = \prod_{i\in N} X_i$ is endowed by a topology, in which the relations R_i are continuous then in the corresponding factor-topology of the space $X|_\sim$ the relations P_i are continuous. The converse statement is also valid.

Now we are turning to the proof of the existence of equilibrium points in the game (5.2).

If the relations R_i are continuous and in the space X there exists a countable and dense w.r.t. the relation R_i set, then there exist continuous utility functions on X preserving the ordering R_i (Debreu, 1954).

If R_i are orderings, then such a set in X exists if and only if the set X is separable in the topology whose basis consists of the open intervals defined by the ordering R_i (Fishburn, 1970).

It is well-known that if X is compact in the topology defined above then it is separable. Therefore we can easily prove the following theorem:

Theorem 5.3 *If in a noncooperative game Γ (2.1) the strategy sets X_i are compact, the relations R_i are orderings and in the corresponding product topology of $X = \prod_{i\in N} X_i$ they are continuous, then there are equilibrium points in the game $\bar{\Gamma}_A$.*

Proof. Consider a game in the payoff function form

$$\Gamma_H = \langle N, \{X_i\}, \{H_i\}_{i\in N}\rangle,$$

in which the payoff functions $H_i(x)$ are continuous in $x \in X$ and preserving the orderings R_i. By Glicksberg's theorem (1952) there are equilibrium points in the mixed extension of Γ_H. It is not difficult to show that they are equilibrium points in the game $\bar{\Gamma}_A$. □

For finite games the converse result is also valid.

Theorem 5.4 (Yanovskaya, 1974) *If the sets X_i are finite for all $i \in N$, then for any equilibrium point $\bar{\mu}$ of the game Γ_A there exists a game in payoff function form $\Gamma_H = \langle N, \{X_i\}_{i\in N}, \{H_i\}_{i\in N}\rangle$ such that $\bar{\mu}$ is an equilibrium point of the mixed extension of Γ_H and for all $i \in N$, $x, y \in X$*

$$\begin{aligned} xP_iy &\iff H_i(x) > H_i(y),\\ xI_iy &\iff H_i(x) = H_i(y). \end{aligned}$$

Example. Consider the ordinal matrix game 2×2

$$\begin{pmatrix} a & c \\ c & b \end{pmatrix}, \tag{5.3}$$

in which the preference relation of the player 1 is the following: $aPbPc$ (the preference relation of the player 2 is opposite). Denote by $(x, 1-x)$, $(y, 1-y)$ mixed strategies of the players 1,2 respectively. Thus, a mixed strategy pair may be denoted by (x, y), $x, y \in [0, 1]$. It is easy to note that equilibrium points in the minimal non-Archimedian extension of this game are (x, y) such that

$$x \in \left(0, \frac{1}{2}\right), y \in \left(0, \frac{1}{2}\right). \tag{5.4}$$

The set of optimal strategies of players in all matrix games with numerical payoffs satisfying the inequalities $a > b > c$ is the set $\{(x, y)\}$, where

$$x = y \in \left(0, \frac{1}{2}\right). \tag{5.5}$$

5.2 Games with arbitrary reflexive preference profiles

For the definition of a mixed extension of such games $\Gamma = \langle N, \{X_i\}_{i\in N}, \{R_i\}_{i\in N}\rangle$ we shall use the definition of the mixed extension of a binary relation given in the third case of Section 4, taking into account that in that case the strict preference relations were considered. It would be also possible to make use the definition (4.8), because the definition has a sense not only for asymmetric relations. However, in the general case, for such extensions the existence theorems turned out well only for finite games.

Thus, let $\sigma-$algebras $\mathfrak{X}_i$ of subsets X_i containing all singletons and the whole sets X_i be given. Let the relations P_i of strict preferences be measurable in the sense that

$$\{x \in X \,|\, xP_iy\{\in \mathfrak{X}, \ \{x \in X \,|\, yP_ix\} \in \mathfrak{X}$$

for all $y \in X, i \in N$, where $\mathfrak{X}$ is the $\sigma-$ algebra of the subsets of $X = \prod_{i\in N} X_i$, generated by the $\sigma-$algebras $\mathfrak{X}_i$. Let M_i be the set of mixed strategies of the player i, i.e. the set of probability measures defined on $\mathfrak{X}_i$, $M = \prod_{i\in N} M_i$.

We let a strategy $n-$tuple $\mu \in M$ be more preferable for a player $i \in N$ than a strategy $n-$tuple ν, if

$$\mu\nu\{(x,y) \mid xP_iy\} > \mu\nu\{(x,y) \mid x\bar{P}_iy\}, \tag{5.6}$$

where $\mu\nu \in M \times M$. Inequality (5.6) means that if we compare the outcomes of the strategy choices according to μ and ν, then the outcomes in first case in the majority of cases are more preferable for the player 1, then other outcomes. This inequality is evidently equivalent to the following one:

$$\mu\nu\{(x,y) \mid xP_iy\} > 1/2. \tag{5.7}$$

Definition (5.7) is extended as follows: let $\alpha_i \in (0,1)$. Then the relations $\tilde{P}_{\alpha_i}$ on M are defined by

$$\mu\tilde{P}_{\alpha_i}y \iff \mu\nu\{(x,y) \mid xP_iy\} > \alpha_i. \tag{5.8}$$

In fact, the relations $\tilde{P}_{\alpha_i}$ are extensions of the relations P_i, since for degenerate measures μ_x, ν_y with supports x, y respectively

$$\mu_x\nu_y\{(x,y) \mid xP_iy\} = \begin{cases} 1, & \text{if } xP_iy, \\ 0, & \text{if } x\bar{P}_iy. \end{cases}$$

The less α_i, the more "weak" is the relation $\tilde{P}_{\alpha_i}$: If $\alpha_{i_1} < \alpha_{i_2}$, then

$$\mu P_{\alpha_{i_2}}\nu \implies \mu P_{\alpha_{i_1}}\nu.$$

The relations $\tilde{P}_{\alpha_i}$ may not possess properties usually imposed on the preference relations: they are neither reflexive, nor anti-reflexive, nor else transitive. It is possible to show that sometimes they are even not acyclic. Nevertheless, it turns out possible to prove the existence of equilibrium points in the games

$$\bar{\Gamma}_{\bar{\alpha}} = \langle N, \{M_i\}_{i\in N}, \{\tilde{R}_{\alpha_i}\}\rangle, \;\; \bar{\alpha} = (\alpha_1, \ldots \alpha_n) \tag{5.9}$$

for some vectors $\bar{\alpha}$, where $\tilde{R}_{\alpha_i}$ are the reflexive relations corresponding to the relations $\tilde{P}_{\alpha_i}$. Of course, it would be interesting to prove the existence of equilibrium points for the games (5.9) for the most minimal values α_i, because if μ^* if an equilibrium point for the game $\bar{\Gamma}_{\bar{\alpha}}$, then it is also an equilibrium point for the game $\bar{\Gamma}_{\bar{\beta}}$, for any vector $\bar{\beta} \geq \bar{\alpha}$.

Let us define the following functions on $X \times X$ for each $i \in N$:

$$K_i(x,y) = \begin{cases} 1, & \text{if } xP_iy, \\ 0, & \text{if } x\bar{P}_iy, \end{cases}.$$

If the set X is finite, then the function $K_i(x,y)$ is defined by the adjacency matrix of the graph G whose set of vertices is X and the edges are defined by the relation P_i. For each $\mu, \nu \in M$ denote

$$K_i(\mu,\nu) = \int_{X \times X} K_i(x,y)\, d\mu(x)\, d\nu(y).$$

Then

$$\mu \tilde{P}_{\alpha_i} \nu \iff K_i(\mu,\nu) > \alpha_i.$$

It is clear that μ^* is an equilibrium point for the game $\bar{\Gamma}_{\bar{\alpha}}$, if for all $i \in N, y_i \in X_i$

$$K_i(\mu^* \| y_i, \mu^*) \leq \alpha_i.$$

For $\mu^j \in \prod_{j \neq i} M_j$ denote by $v_i(\mu^i)$ the value of zero-sum two-person game Γ_{μ^i} on $X_i \times X_i$ with the payoff function $K_i(\mu^i \| x_i, \mu^i \| y_i)$, if it exists.

Theorem 5.5 (Yanovskaya, 1973) *If X_i are compact Hausdorff spaces, the relations P_i on X are such that the sets $\{(x^i, x_i), y) \mid xP_iy\}$ are open in $X^i \times X$ for all $x_i \in X_i$, $i \in N$, then there are equilibrium points in the game $\bar{\Gamma}_{\bar{\alpha}}$, where $\alpha_i = \sup_{\mu^i} v_i(\mu^i)$.*

In the proof of the Theorem the existence of the values $v_i(\mu^i)$ of the corresponding zero-sum two-person games is also established.

Corollary 5.6 *If in the conditions of Theorem 5.5 the relations P_i on X are asymmetric, then $\alpha_i \leq 1/2$.*

The following example shows that this estimation for α_i is exact.

Example 1. Consider an one-person game on the unit interval, in which the preference relation of the player R is defined by

$$\begin{aligned} &xPy, && \text{if } y > x + \frac{1}{2} \text{ or } x > y > x - \frac{1}{2}, \\ &xIy && \text{otherwise.} \end{aligned}$$

The relation P satisfies the conditions of Theorem 5.5 and asymmetric. The adjacency function of this relation $K(x,y)$ is defined on the unit square and has the form

$$K(x,y) = \begin{cases} 1, & \text{if } y > x + \frac{1}{2} \text{ or } x > y > x - \frac{1}{2}, \\ 0 & \text{otherwise.} \end{cases}$$

The value of the game on the unit square with the payoff function $K(x,y)$ is equal to 1/2. It is easy to note that for $\alpha < 1/2$ there are no equilibrium points in the games $\bar{\Gamma}_\alpha$. The optimal strategy of both players in the game $\bar{\Gamma}_{1/2}$ is the uniform distribution on the interval $[0,1]$.

Example 2. Let Γ be again an one-person game with three strategies leading to the outcomes a, b, c and the preference relation P of the player is defined by $aPbPcPbPaPcPa$. It is symmetric, but non-reflexive. For any mixed strategies μ, ν we have

$$\mu\nu\{(x,y) \,|\, xPy,\ x,y \in \{a,b,c\}\} = 1 - \mu(a)\nu(a) - \mu(b)\nu(b) - \mu(c)\nu(c),$$

and, thus, the strategy $\mu : \mu(a) = \mu(b) = \mu(c) = 1/3$ is an equilibrium if $\alpha = 2/3$, and there are no equilibria for $\alpha < 2/3$.

This game may be represented as a social choice problem with three individuals A, B, C and three alternatives a, b, c where the profile of individual preferences (P_A, P_B, P_C) is the following:

$$\begin{array}{lll} aP_Ab, & aP_Ac, & bI_Ac, \\ bP_Ba, & bP_Bc, & zI_Bc, \\ cP_Ca, & cP_Cb, & aI_Cb, \end{array}$$

where I_A, I_B, I_C are the corresponding equivalence relations. As the problem is symmetric, any anonymous social choice rule should choose all the alternatives. However if the random choices of outcomes are possible there is the anonymous social choice rule prescribing to choose all outcomes equiprobably.

Example 3. Let us consider one more the example of 2×2 matrix game (5.3):

$$A = \|a_{ij}\| = \begin{pmatrix} a & c \\ c & b \end{pmatrix}, \tag{5.10}$$

where the preference relation P_1 of the player 1 is aP_1bP_1c, aP_1c. Therefore, the relation R_1 corresponding to P_1 is an ordering. By the definition the preference relation of player 2 is converse to that of player 1: $cP_2P_2bP_2a$, cP_2a.

If a mixed extension of the game satisfies all the axioms of the expected utility theory besides, possibly, the Archimedian axiom, then there exists a finite number of utility functions such that the lexicographic relation between their expected values is equivalent to the preference relation on the pairs of mixed strategies (Hausner, 1954, or Birkhoff, 1948). For this example one of such matrices is the following:

$$\begin{pmatrix} 1,0 & 0,0 \\ 0,0 & 0,1 \end{pmatrix}.$$

As it had been shown in Fishburn (1972) in this game there are no optimal mixed strategies of the players. Let us find its optimal strategies of the game (5.10) in the mixed extension $\bar{\Gamma}_{\bar{\alpha}}$ for some $\bar{\alpha} = (\alpha_1, \alpha_2)$. Denote the optimal strategies by the players by $\mu^* = (p, 1-p)$, $\nu^* = (q, 1-q)$ respectively. Then by the definition of the equilibrium points in $\bar{\Gamma}_{\bar{\alpha}}$ the following inequalities should hold:

$$\begin{aligned} &(i\nu^*)(\mu^*\nu^*)\{(a_{ik}, a_{il}) \,|\, a_{ik}P_1a_{il}\} \leq \alpha_1, \quad i,j,k,l = 1,2,\ k \neq l \\ &(\mu^*j)(\mu^*\nu^*)\{(a_{kj}, a_{lj}) \,|\, a_{kj}P_2a_{lj}\} \leq \alpha_2. \end{aligned} \tag{5.11}$$

Inequalities (5.11) are equivalent to the following ones:

$$\begin{aligned} q(1-pq) &\leq \alpha_1, \\ (1-q)[p(1-q) + q(1-p)] &\leq \alpha_1, \\ (1-p)[(1-p)(1-q) + pq] &\leq \alpha_2, \\ p[(1-p)(1-q) + pq] &\leq \alpha_2, \\ p, q \in [0,1], 0 < \alpha_1, \alpha_2 \leq 1/2.& \end{aligned} \tag{5.12}$$

System (5.12) has many solutions, in particular, for $\bar{\alpha} = (3/8, 3/8)$ the optimal strategies in the game $\bar{\Gamma}_{\bar{\alpha}}$ are defined by $p = q = 1/2$. However there exist optimal strategies in the games $\bar{\Gamma}_{\bar{\alpha}}$ where α_i, $i = 1, 2$ are a little less than 3/8, in this case $p < q$, $q > 1/2$.

The last example illustrates the difference of two approaches to the definition of preference relations on mixed strategies $n-$tuples: the expected utility approach and the one given above. In fact, in matrix game (5.10), where $a > b > c$ are some numbers, the optimal strategies of the players $(x^*, 1-x^*)$, $(y^*, 1-y^*)$ satisfy the inequalities $0 < x^* = y^* < 1/2$.

Return now to the mixed extensions of players' preference relations defined in (4.8). Denote them by Q_i, i.e.

$$\mu Q_i \nu \iff \mu\nu\{(x,y) \,|\, xP_iy\} > \mu\nu\{(x,y) \,|\, yP_ix\}, \quad x, y \in X,\ \mu, \nu \in M.$$

Consider the mixed extension of Γ defined by the relations Q_i :

$$\bar{\Gamma}_Q = \langle N, \{M_i\}_{i \in N}, \{\tilde{Q}_i\}_{i \in N} \rangle,$$

where $\tilde{Q}_i$ are the reflexive relations corresponding to the relations Q_i.

Theorem 5.7 *If in the game Γ the strategy sets of the players are finite, then in the game $\bar{\Gamma}_Q$ there are equilibrium points.*

The proof is fulfilled with help of a fixed point theorem similarly to that of Theorem 5.5 with only difference that instead of the functions $K_i(x, y)$, defined on $X \times X$ here we should consider the following functions

$$\tilde{K}_i(x, y) = \begin{cases} 1, & \text{if } xP_iy, \\ -1, & \text{if } yP_ix, \\ 0 & \text{otherwise.} \end{cases} \tag{5.13}$$

Remark. The extension of Theorem 5.7 to the case of infinite sets X_i is possible only in special cases. In fact, since the relations P_i are non reflexive, the functions $\tilde{K}_i(x, y)$ (5.13) do not satisfy the conditions of minimax theorems.

Example 4. Let us find equilibrium points in the mixed extension $\bar{\Gamma}_Q$ of the matrix game considered in Example 3. If in its extension $\bar{\Gamma}_Q$ there are equilibrium points $\mu^* = (p, 1-p), \nu^* = (q, 1-q)$, then they should satisfy the inequalities

$$\begin{aligned} 0 \ &\geq \ (i\nu^*)(\mu^*\nu^*)\{(a_{ik}, a_{il}) \,|\, a_{ik}P_1a_{il}\} \\ &\quad -(i\nu^*)(\mu^*\nu^*)\{(a_{ik}, a_{il}) \,|\, a_{il}P_1a_{ik}\}, \\ 0 \ &\geq \ (\mu^*j)(\mu^*\nu^*)\{(a_{kj}, a_{ij}) \,|\, a_{kj}P_2a_{lj}\} \\ &\quad -(\mu^*j)(\mu^*\nu^*)\{(a_{kj}, a_{lj}) \,|\, a_{lj}P_2a_{kj}\}, \\ &\quad i, j = 1, 2;\ k, l = 1, 2;\ k \neq l \end{aligned} \tag{5.14}$$

Rewriting these inequalities for $i = 1, 2$, $j = 1, 2$ we obtain

$$\begin{aligned} &i = 1 : q(1-pq) - (1-q)(pq + (1-p)(1-q) \leq 0, \\ &i = 2 : (1-q)((1-p)q + p(1-q) - pq) - q(pq + (10p)(1-q)) \leq 0, \\ &j = 1 : (1-p)(pq + (1-p)(1-q)) - p(1-pq) \leq 0, \\ &j = 2 : p(pq + (1-p)(1-q))) - (1-p)(p(10q) - q(1-p)) \leq 0. \end{aligned} \tag{5.15}$$

The unique solution of the system (5.15) is

$$p = q = \frac{3 - \sqrt{5}}{2},$$

i.e. both players choose their strategies with the probabilities equal to the gold section.

Acknowledgement
This chapter is the author's translation of E.B. Yanovskaya, "Situatsii ravnovesia v obshchikh beskoalotsionnykh igrakh i ikh smeshannykh rasshireniyakh," in: "Teoretiko-igrovye voprosy prinyatiya reshenij" (Game-theoretic problems of decision making), Leningrad, "Nauka", 1978, 43–65 (in Russian).

References

1. Birkhoff, G. (1948): *Lattice theory.* American Mathematical Society Publications, XXV, New York.

2. Debreu, G. (1954): "Representation of a preference ordering by a numerical function," in: Decision processes, R.M.Thrall, C.H. Coombs, R.L.Davis (Eds) New York, Wiley, 159–165.

3. Eilenberg, S. (1941): "Ordered topological spaces," American Journal of Mathematics, 63, 39–45.

4. Farquharson, R. (1955): "Sur une généralisation de la notion d'équilibrium," C.r. Acad.Sci., Paris, 240, N1, 46–48.

5. Fishburn, P. (1970): *Utility theory for decision making.* New York, Wiley.

6. Fishburn, P. (1972): "Lotteries and social choices," *Journal of Economic Theory*, 5, 189–207.

7. Fishburn, P. (1972): "On the foundation of game theory," *International Journal of Game Theory*, 1, 65–71.

8. Gillies, D. (1959): "Solutions of general non-zero-sum games," in: Contr. to the theory of games, v.4, Princeton, Princeton Univ. Press (Ann.Mathem. Stud., N 39) 47–85.

9. Glicksberg, I.L. (1952): "A further generalization of the Kakutani fixed point theorem, with application to Nash equilibrium points," Proc. Amer. Math. Soc., 3, 170–174.

10. Hausner, M. (1954): "Multidimensional utilities," in: Decision Processes, Wiley, New York, 167–180.

11. Kulakovskaya, T. (1976): "Classical optimality principles in infinite cooperative games," in: "Sovremennye napravlenia teorii igr" (Modern tendencies of game theory). E.Vilkas (Eds), Vilnius, Mintis, 94–108 (in Russian).

12. Nash, J. (1951): "Non-cooperative games," Ann. Math., 54, 286–295.

13. von Neumann, J., and O. Morgenstern (1944): *Theory of Games and Economic Behavior.* Princeton: Princeton Unversity Press.

14. Nikaido, H., and K. Isoda (1955): "Note on noncooperative convex games," *Pacific Journal of Mathematics*, 5, suppl.1, 807–815.

15. Walsh, J.E. (1969: "Discrete two person game theory with median payoff criterion," Opsearch, 6, N1, 83–97.

16. Yanovskaya, E. (1973): "On a mixed extension of a general noncooperative game," in: "Teoretiko-igrovye voprosy prinyatia reshenij" (Game-theoretic problems of decision-making). Central Institute of Economics and Mathematics of Academy of Sciences of the USSR, 213–227 (in Russian).

17. Yanovskaya, E. (1974): "Equilibrium points in games with non-archimedian utilities," in: "Mathematical methods in social sciences". Institute of Physics and Mathematics of Lituanien Acad.Sci., Vilnius, 4, 98–118 (in Russian).

18. Yanovskaya, E. (1976): "The mixed extension of a binary relation," in: "Mathematical methods in social sciences". Institute of Physics and Mathematics of Lituanien Acad.Sci., Vilnius, 6, 152–166 (in Russian).

Chapter 4

On the theory of optimality principles for noncooperative games

V. Lapitsky

Abstract: The category structure on the class of noncooperative games is introduced to study universal properties of equilibrium and some other optimality principles. It is proved that the generalized equilibrium is a functor from this category to the category of sets. Generalizing this fact, an axiomatic definition of optimality principle is proposed. Some more optimality principles are introduced and studied.

Key words: Noncooperative game, category of games, generalized equilibrium, optimal functor, axiomatics of optimality

1 Introduction

The notion of optimality principle is one of the basic concepts of modern game theory. According to N. Vorob'ev (1994), it is possible to say that the general line of development of this theory consists, in a sense, in formalizing different concepts of optimality by some principles and in elaborating methods of their realization: both theoretical (theorems of existence) and practical (explicit procedures for calculating) ones. But despite the empirical analysis of numerous different ideas of optimality, at the moment there is no general theory or even definition of optimality principle – especially for noncooperative games.

In this paper we propose a general definition of optimality principle for the class of noncooperative games with the help of natural category structure on it. To be more precise, we introduce a (quite natural) categorial structure on the class of noncooperative games over an arbitrary category and then, using the fact that the generalized equilibrium is a functor from this category, we define an optimality principle as a suitable functor from this category of games to the category of sets. For the first time this idea (for the particular case of equilibrium points) was proposed by N. Vorob'ev; apparently similar considerations were used in an absolutely different context by Takahara *et al.* (1980, 1981) for games against the Nature.

The paper is organized as follows. In Section 2 we introduce the general categorial formalism and related basic notions (for definitions and some general results of the theory of categories see Gelfand and Manin, 1997, or MacLane, 1971). In Section 3 we study categorial properties of several kinds of equilibrium and prove that the so called $\mathfrak{A}$-equilibrium is a functor from the corresponding category. Using an axiomatic approach, in Section 4 we introduce our general definition of optimality principle and give some examples of such principles. In Section 5 an additional axiom characterizing an equilibrium is introduced.

2 Categories and games

2.1 Categories of games

Considering the classical definition of a noncooperative game (see Vorob'ev, 1994) one can see that it can be easily reformulated in purely categorical terms. Namely we have

Definition 2.1 A noncooperative game in a category C is a collection

$$\Gamma = \langle I, \{S_i\}, \{K_i\}, \{\varphi_i\}_{i\in I}\rangle,$$

where I is a set (of players), $S_i \in \mathrm{Ob}C$ is a "strategic object" of i-th player, $K_i \in \mathrm{Ob}C$ is a "payoff object" of i-th player, $\varphi_i : S = \prod_{j\in I} S_j \to K_i$ is a morphism of category C (payoff morphism of i-th player).

Here some remarks are needed:

(1) Naming S_i and K_i "strategic" and "payoff" objects, we do not mean that they consist of some "strategies" or "payoffs". Being objects of a category, they have neither points (elements) nor an inner structure.

(2) We do not suppose that C possesses direct products but claim it only for the family of objects $\{S_i\}_{i \in I}$.

(3) Not being a mapping, the morphism φ_i does not map the "outcomes" of $\mathfrak{A}$ to the "payoffs".

From the categorical point of view the same construction can be defined in a different way. We call a *game I-scheme* the following diagram scheme (or category):

$$\Delta_I = \{S_i \xleftarrow{p_i} S \xrightarrow{\varphi_i} K_i\}_{i \in I}.$$

For an arbitrary category C, we can consider a category C^{Δ_I} of game I-schemes (or functors) with values in C (see Grotendieck, 1957; MacLane, 1971). Let $\Delta_I(C)$ denote the full subcategory of those diagrams in which the family $\{p_i\}_{i \in I}$ determines a representation of S as a direct product of the family of objects $\{S_i\}_{i \in I}$. It is clear that the objects of this category are exactly the games from the definition above, but now they automatically inherit the category structure from C.

Let us rewrite this structure explicitly.

Definition 2.2 Let $\Gamma = \langle I, \{S_i\}, \{K_i\}, \{\varphi_i\}_{i \in I}\rangle$ and $\Gamma' = \langle I, \{S'_i\}, \{K'_i\}, \{\varphi'_i\}_{i \in I}\rangle$. Then a morphism $\Gamma \to \Gamma'$ of category $\Delta_I(C)$ is a family $(f_i, f^i)_{i \in I}$, where $f_i : S_i \to S'_i$, and $f^i : K_i \to K'_i$, such that the diagram

$$\begin{array}{ccc} S = \prod S_j & \xrightarrow{f = \prod f_i} & S' = \prod S'_j \\ \downarrow \varphi_i & & \downarrow \varphi'_i \\ K_i & \xrightarrow{f^i} & K'_i \end{array}$$

is commutative for all i.

A natural composition of such morphisms yields indeed a category structure on $\Delta_I(C)$.

The main properties of this category were studied in Lapitsky (1983). Now we consider one more general case rejecting the previous limitation: the fixed set of players.

Let

$$\begin{aligned} \Gamma &= \langle I, \{S_i\}, \{K_i\}, \{\varphi_i\}_{i \in I}\rangle, \\ \Gamma' &= \langle J, \{T_j\}, \{L_j\}, \{\psi_j\}_{j \in J}\rangle \end{aligned}$$

be two games from categories $\Delta_I(C)$ and $\Delta_J(C)$ respectively.

Definition 2.3 A morphism from Γ to Γ' is a collection

$$(F, \{f_i\}, \{f^i\}_{i \in I}),$$

where

$F : I \to J$ is a mapping of sets,

$f_i : T_{F(i)} \to S_i$ is a morphism of category C,

$f^i : L_{F(i)} \to K_i$ is a morphism of category C.

The only condition required from this collection is that all diagrams

$$\begin{array}{ccc} S & \xleftarrow{f} & T \\ \downarrow \varphi_i & & \downarrow \psi_{F(i)} \\ K_i & \xleftarrow{f^i} & L_{F(i)} \end{array}$$

are commutative. Here f is naturally induced by $\{f_i\}$ and can be formalized by $f = (\prod_{j \in F(i)} f_{(j)}) \circ \mathrm{pr}$, where pr is the canonical projection of T onto $\prod_{j \in F(i)} T_j$, and $f_{(j)} = \underset{i \in F^{-1}(j)}{(\times)} f_i : T_j \to \prod_{i \in F^{-1}(j)} S_i$.

Clearly such morphisms (changing the number of players) determine a category structure on the class of all noncooperative games, i.e. on the class $\bigcup_I \Delta_I(C)$; we denote this category by $\Delta(C)$, and the category $\Delta(\mathbf{Set})$ by **gam**. Restricting ourselves by the morphisms with $F = \mathrm{id}$, we obtain previous categories $\Delta_I(\mathcal{C})$ with different sets I and with reversed arrows, i.e., the categories $\Delta_I(C)^o$ are subcategories (not full!) of $\Delta(C)$.

The morphisms of this category are characterized by the following theorems.

Theorem 2.4 *Assume that $\mathcal{C}$ satisfies the axiom AB4* of Grothendieck (1975). Then the morphism $\overline{f} : \Gamma \to \Gamma'$ is a monomorphism iff F is injective and all f^i and f_i are epimorphic.*

Theorem 2.5 *Assume that $\mathcal{C}$ has an initial object. Then the morphism $\overline{f} : \Gamma \to \Gamma'$ is an epimorphism iff F is surjective and all $f^{(j)}$ and $f_{(j)}$ are monomorphic.*

Proofs of these theorems can be found in Lapitsky (1983).

2.2 Games in categories

Continuing the remarks of Subsection 2.1, we can say that the game Γ represents a scheme of rules or "archetype" of some game in common sense, although it can not be played itself. What would be then the meaning of "playing a game" from $\Delta_I(C)$? A usual categorical tool to introduce a structure on objects of a category is a point functor (cf. Gelfand and Manin, 1997). We define a functor

$$\Phi_C^I : C \times \Delta_I(C) \rightsquigarrow \mathbf{gam}_I = \Delta_I(\mathbf{Set})$$

called the *functor of realization*, by

$$\Phi_C^I(A, \Gamma) = \Gamma(A) = \langle I, \{\mathrm{Hom}(A, S_i)\}, \{\mathrm{Hom}(A, K_i)\}, \{\mathrm{Hom}(A, \varphi_i)\}_{i \in I} \rangle.$$

Here the components of the morphism $\Phi_C^I(\alpha, \beta)$ are equal to $\beta_i \chi_i \alpha$ where $\alpha : A \to B$, and χ lies in $\mathrm{Hom}(B, S_i)$ or $\mathrm{Hom}(B, K_i)$, β_i is the corresponding component of the morphism $\beta : \Gamma \to \Gamma'$. It is easy to see that Φ_C^I is indeed a functor which is contravariant in the first variable and covariant in the second one. For every $A \in \mathrm{Ob}C$, we can speak of A-realization of Γ - the usual (i.e. having strategies and payoffs) game $\Gamma(A) \in \mathrm{Ob}\,\mathbf{gam}_I$. Thus every game from $\Delta_I(C)$ determines a (contravariant) functor from C to $\mathbf{gam}_I$ but not vice versa. Denote the category $\mathrm{Funct}^0(C, \mathbf{gam}_I)$ by ${\Delta_I}^1(C)$, the category $\mathrm{Funct}^o(C, \mathbf{Set})$ by $\overline{C}$.

Then we have:

Proposition 2.6 *(1)* ${\Delta_I}^1(C) = \Delta_I(\overline{C})$;

(2) $\Delta_I(C)$ *is a full subcategory in* ${\Delta_I}^1(C)$;

(3) restriction of the functor $\Phi_{\overline{C}}^I$ *on* $C \times \Delta_I(C)$ *coincides with* Φ_C^I.

Proof.

(1) Let $\Gamma \in \mathrm{Ob}{\Delta_I}^1(C)$ i.e. $\Gamma : C \rightsquigarrow \mathbf{gam}_I$. Then we can consider a functor $\Gamma_{S_i} : C \rightsquigarrow \mathbf{Set}$ sending object A of category C to the set $S_i(A)$ of the game $\Gamma(A)$, and a functor Γ_{K_i} sending A to the set $K_i(A)$ of the same game. For every $i \in I$ we have also a functor morphism $\Gamma_{\varphi_i} : \Gamma_S \to \Gamma_{K_i}$ that assigns to an object A the mapping $\varphi_i(A)$ in the game $\Gamma(A)$. Therefore we can associate with an arbitrary $\Gamma \in \mathrm{Ob}{\Delta_I}^1(C)$ a diagram with the scheme Δ_I of contravariant functors from C to $\mathbf{Set}$, i.e. an object of $\Delta_I(\overline{C})$. Vice versa, every diagram from $\Delta_I(\overline{C})$ can be viewed as a functor from C to $\mathbf{gam}_I$.

Consider now the morphisms of the category ${\Delta_I}^1(C)$. We can naturally associate with every functor morphism $\Gamma \to \Gamma'$ two families of functor morphisms $\Gamma_{S_i} \to \Gamma'_{S_i}$ and $\Gamma_{K_i} \to \Gamma'_{K_i}$, and vice versa, every family $\{\Gamma_{S_i} \to \Gamma'_{S_i}, \Gamma_{K_i} \to \Gamma'_{K_i}\}$ determines a morphism $\Gamma \to \Gamma'$; we have to check only the coincidence of commutation conditions. In both cases they have the following form:

$$\begin{array}{ccc} \Gamma_S(A) & \longrightarrow & \Gamma_S(A)' \\ \downarrow & & \downarrow \\ \Gamma_{K_i}(A) & \longrightarrow & \Gamma_{K_i'}(A) \end{array}$$

(2) The imbedding of $\Delta_I(C)$ in ${\Delta_I}'(C)$ is determined by the functor $\Phi^I_C(\circ, \Gamma)$. Using a), we can conclude that the category $\Delta_I(C)$ consists of diagrams in $\Delta_I(\overline{C})$ with representable functors Γ_{S_i} and Γ_{K_i}. Therefore the fullness of $\Delta_I(C)$ in ${\Delta_I}^1(C)$ follows now from the classical theorem on the fullness of functor Hom_C (see Gelfand and Manin, 1997).

(3) This statement follows immediately from the same theorem. □

Corollary 2.7 *(1)* $\Delta^1(C) = \Delta(\overline{C})$*;*

(2) $\Delta(C)$ *is a full subcategory in* $\Delta^1(C)$*;*

(3) restriction of the functor $\Phi_{\overline{C}}$ *on* $C \times \Delta(C)$ *coincides with* Φ_C*.*

The same game can therefore be played in different realizations, and we can consider A-strategies, A-payoffs and A-payoff functions. The role of a pure realization (i.e. pure strategies and so on) is played here by z-realization, where z is a terminal object (right zero) of C (i.e. such an object that the set $\mathrm{Hom}(X, z)$ has a unique element for every $X \in \mathrm{Ob}C$). The terminal object does not necessarily exist in the category C, but it always exists in $\overline{C}$: it is a functor sending every object to a one-point set. Thus we can consider a pure realization of Γ in the larger category $\Delta^1(C)$.

2.3 Preferences in categories

The playing mechanism introduced in the previous section does not, however, allow us to consider optimality in games in categories if the players do not have any preferences. Therefore we need to introduce some order structure on the objects K_i mentioned in the definition of a game. For this purpose we use once more the traditional tool: the point functor (see Gelfand and Manin, 1997).

Let $h_X = \mathrm{Hom}(\circ, X)$ be a functor of points of object X (or functor represented by X). We say that an object A of the category C is *ordered* if there exists a functor $\vartheta : C \rightsquigarrow$ **OrdSet** that makes the following diagram commutative:

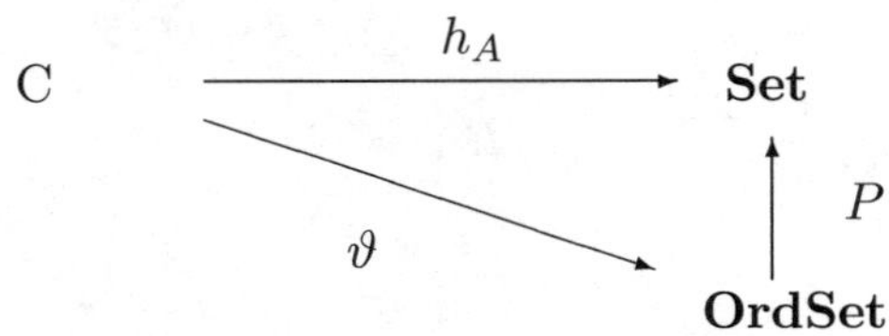

where P is a forgetful functor. In other words this means that all sets $\mathrm{Hom}(X, A)$ become partially ordered in a consistent way. Let us introduce a new category Ord$C$ (the category of orders in C). Its objects are pairs (A, ϑ), where $A \in \mathrm{Ob}C$, and ϑ is an order on A; its morphism $\tilde{f} : (A, \vartheta_A) \to (B, \vartheta_B)$ is a pair (f, φ), $f : A \to B$, and $\varphi : \vartheta_A \to \vartheta_B$ is a functor morphism and $\varphi(X) = h^X(f)$ for every $X \in \mathrm{Ob}C$. The functor $A \rightsquigarrow (A, h_A)$ makes the category C full in Ord$C$.

Like other structures on the objects of a category, the order can be introduced by means of diagrams. Let us select a subobject R of A and a morphism of inclusion $r : R \hookrightarrow A \times A$ such that the following conditions hold:

(i) There is a cartesian square

$$\begin{array}{ccc} \Delta & \longrightarrow & R \\ \downarrow & & \downarrow \overline{r} = r\sigma \\ R & \xrightarrow{r} & A \times A \end{array}$$

where $\sigma : A \times A \to A \times A$ changes the order of components, Δ is the diagonal.

(ii) Let

$$U = (R \times A) \underset{A \times A \times A}{\times} (A \times R),$$

and $p : A \times A \times A \to A \times A$ be a projection on the first and third components. Then $U \to A \times A \times A \xrightarrow{P} A \times A$ factors through R, i.e. there exists a commutative diagram

$$\begin{array}{ccccc} \mathrm{U} & \longrightarrow & A \times A \times A & \xrightarrow{p} & A \times A \\ & \searrow & & \nearrow_{r} & \\ & & R & & \end{array}$$

Proposition 2.8 *Define an order on an arbitrary set* $\mathrm{Hom}(X, A)$ *by*

$$f \leq g \Longleftrightarrow \quad \textit{there exists a commutative diagram}$$

$$\begin{array}{ccc} X \times X & \xrightarrow{f \times g} & A \times A \\ \searrow & & \nearrow \\ & R & \end{array}$$

Then this order makes an object A ordered in C.

Proposition 2.9 *A morphism $\varphi : A \to B$ of two ordered in this way objects is monotone if and only if the morphism $R_A \to A \times A \xrightarrow{\varphi \times \varphi} B \times B$ factors through R_B.*

Proofs of these statements are straightforward and we omit them.

In the sequel we suppose that the objects K_i of any game Γ are ordered and their morphisms in the definition of the morphisms of games are monotone.

3 Equilibrium in categories

Now we can introduce in the category $\Delta(C)$ a notion of (Nash) equilibrium. First of all we define it for the category **gam**.

Definition 3.1 An outcome $s \in S$ of a game $\Gamma \in \mathrm{Ob}\mathbf{gam}$ is called a weak equilibrium point if $\varphi_i(s||s_i) \not> \varphi_i(s)$ for every $i \in I$ and every $s_i \in S_i$.

An outcome s is called an equilibrium point if $\varphi(s_i||s_i) \leq \varphi_i(s)$ for every $i \in I$ and every $s_i \in S_i$.

As for the category $\Delta(C)$, we define an equilibrium in a game Γ as an equilibrium in A-realization of this game. Namely:

Definition 3.2 The morphism $s : A \to S$ is called a weak equilibrium A-point if for every $i \in I$ and every $s_i : A \to S_i$

$$\varphi_i(s_i||s_i) \not> \varphi_i s\,.$$

A morphism $s : A \to S$ is called an equilibrium A-point if for every $i \in I$ and every $s_i : A \to S_i$

$$\varphi_i(s_i||s_i) \leq \varphi_i s\,.$$

Consider now categorical properties of equilibrium in the category $\Delta(\mathcal{C})$.

Lemma 3.3 *Let* $\Gamma, \Gamma' \in \mathrm{Ob}\mathbf{gam}$ *and* $(F, \{f_i\}, \{f^i\}_{i\in I}) : \Gamma \to \Gamma'$. *Then*

(1) If all f_i are surjective then the image of an equilibrium point under f is an equilibrium point.

(2) If F is surjective and all $f^{(j)}$ are injective then the inverse image of a weak equilibrium point (if any) is a weak equilibrium point.

Proof. (1) Let $t^* \in T$ be an equilibrium point and $s_i \in S_i$ be a strategy such that $\varphi_i(f(t^*)||s_i) \not\leq \varphi_i(f(t^*))$. By the surjectivity of f_i there exists $t_{F(i)} \in T_{F(i)}$ such that $f_i(t_{F(i)}) = s_i$. Since t^* is an equilibrium point, we have $\psi_{F(i)}(t^*||t_{F(i)}) \leq \psi_{F(i)}(t^*)$. Using the monotonicity of f^i, we have $f^i\psi_{F(i)}(t^*||t_{F(i)}) \leq f^i(\psi_{F(i)}(t^*))$. Since (F, f_i, f^i) is a morphism of category **gam**, we have $f^i\psi_{F(i)}(t) = \varphi_i f(t)$ for every $t \in T$. Hence $\varphi_i f(t^*||T_{F(i)}) \leq \varphi_i f(t^*)$ and $\varphi_i(f(t^*)||s_i) \leq \varphi_i f(t^*)$, i.e. we obtain a contradiction.

(2) Let s^* be a weak equilibrium point in the game Γ and $f(t^*) = s^*$. We show that t^* is also a weak equilibrium point. Suppose that there exists $t_j \in T_j$ such that $\psi_j(t^*) < \psi_j(t^*||t_j)$. Since $f^{(j)}$ is injective, there exists $i \in I$ such that $F(i) = j$ and $f^i\psi_j(t^*) \neq f^i\psi_j(t^*||t_j)$. Then by monotonicity of f^i we have $f^i\psi_j(t^*) < f^i\psi_j(t^*||t_j)$. Therefore $\varphi_i f(t^*) < \varphi_i(f(t^*||t_j))$, i.e. $\varphi_i f(t^*) < \varphi_i(f(t^*)||s_i)$ where $s_i = f_i(t_j)$. Hence $\varphi_i(s^*) < \varphi_i(s^*||s_i)$ - a contradiction. □

A similar result on "quasi-functorness" of weak equilibrium points holds also for the category $\Delta(C)$. Denote the set of weak equilibrium points in the A-realization of $\Gamma \in \mathrm{Ob}\Delta(C)$ by $\mathrm{eq}(A, \Gamma)$.

Theorem 3.4 *Assume that C has an initial object. Then every pair of epimorphisms $\alpha \in \mathrm{Hom}_C(A,B)$ and $\beta \in \mathrm{Hom}_{\Delta(C)}(\Gamma,\Gamma')$ determines an inclusion* $\mathrm{eq}(A,\Gamma) \underset{S(A)}{\times} T(B) \hookrightarrow \mathrm{eq}(B,\Gamma')$ *induced by* $\Phi_C(\alpha,\beta)$.

Proof. As usual it suffices to prove this statement for pairs $(\alpha, 1_\Gamma)$ and $(1_A,\beta)$. In the first case apply Lemma 3.3 to the diagrams

$$\begin{array}{ccc} \mathrm{Hom}(B,S) & \longrightarrow & \mathrm{Hom}(A,S) \\ \downarrow & & \downarrow \\ \mathrm{Hom}(B,K_i) & \longrightarrow & \mathrm{Hom}(A,K_i) \end{array}$$

where the horizontal arrows are injective since α is epimorphic.

For the second case, observe that by Theorem 2.5 $f^{(j)}$ are monomorphisms and F is surjective. Now we can apply Lemma 3.3 to the diagrams

$$\begin{array}{ccc} \mathrm{Hom}(A,S) & \overset{f(A)}{\longleftarrow} & \mathrm{Hom}(A,T) \\ \downarrow \prod \varphi_i(A) & & \downarrow \psi_j(A) \\ \prod_{i\in F^{-1}(j)} \mathrm{Hom}(A,K_i) & \overset{f^{(j)}(A)}{\longleftarrow} & \mathrm{Hom}(A,L_j) \end{array}$$

where the arrows below are injective as $f^{(j)}$ are monomorphisms. □

Now we introduce a new principle of optimality that extends the usual equilibrium points even in the classical case. For this purpose we define a new notion connected with the possibilities of the players in a game.

Let Γ be a game from the category $\Delta(C)$.

Definition 3.5 A behaviorial type in a game Γ is a collection $\mathfrak{A} = \{\mathfrak{A}_i\}_{i\in I}$, where $\mathfrak{A}_i \subset \mathrm{Hom}(S_i,S_i)$, such that

(i) $1_{S_i} \in \mathfrak{A}_i$

(ii) $f,g \in \mathfrak{A} \Rightarrow f\,g \in \mathfrak{A}_i$.

Let $A = \prod_{j\in I} A_i$ and $\alpha_i \in \mathrm{Hom}(A_i,A_i)$. We define the morphism $\bar{\alpha}_i : A \to A$ by

$$\bar{\alpha}_i = \prod_{j\in I} a_j \quad \text{where} \quad a_j = \begin{cases} 1_{A_j} & j \neq i \\ \alpha_i & j = i. \end{cases}$$

For $P \subset \mathrm{Hom}(A_i,A_i)$ denote by $\bar{P}$ a set $\{\bar{\alpha}/\alpha \in P\}$.

Definition 3.6 An outcome $s \in \mathrm{Hom}(A, S)$ of the game Γ is said to be an $\mathfrak{A}$-equilibrium point if for every player $i \in I$ and every $\alpha_i \in \mathfrak{A}_i$

$$\varphi_i s \geq \varphi_i \bar{\alpha}_i s.$$

$\mathfrak{A}$-equilibrium points differ from the usual equilibrium points: namely, the players can not change their strategies arbitrarily but should follow some rules. The standard equilibrium is a special case of $\mathfrak{A}$-equilibrium corresponding, for example, to $\mathfrak{A}_i = \mathrm{Hom}(S_i, S_i)$ (or, more conventionally, to $\mathfrak{A}_i$ consisting of all projections to one point).

Up to now we considered an isolated game Γ. Let us return now to the category $\Delta(C)$. Consider a class of pairs $(\Gamma, \mathfrak{A})$ where Γ is a game from the category $\Delta(C)$ and $\mathfrak{A}$ is a behaviorial type in it. These pairs are the objects of a new category where the morphism from one pair to another is a morphism of games $\Gamma \to \Gamma'$ *consistent* with behaviorial types. More precisely, let $g : A \to B$ be a morphism of category C. We denote by g^* and g_* the mappings

$$g^* : \mathrm{Hom}(A, A) \to \mathrm{Hom}(A, B)$$

and

$$g_* : \mathrm{Hom}(B, B) \to \mathrm{Hom}(A, B)$$

induced by

$$\begin{aligned} g^*(\alpha) &= g\alpha, \\ g_*(\beta) &= \beta g. \end{aligned}$$

Definition 3.7 A morphism $(F, \{f_i\}, \{f^i\}_{i \in I}) : \Gamma \to \Gamma'$ where Γ and Γ' have behaviorial types $\mathfrak{A}$ and $\mathfrak{A}'$ respectively is called consistent with these types if for every $i \in I$ we have

$$f^*_{F(i)}(\mathfrak{A}'_{F(i)}) \supset f_{(F(i))^*}(\bar{\mathfrak{A}}_i).$$

We denote this category by $\Delta E(C)$ and the category $\Delta E(\mathbf{Set})$ by **Gam**. It is easy to prove that the functor of realization preserves consistency. Namely we have

Proposition 3.8 *The functor of realization* $\Phi_C : C \times \Delta(C) \rightsquigarrow \mathbf{gam}$ *induces a functor*

$$\bar{\Phi}_C : C \times \Delta E(C) \rightsquigarrow \mathbf{Gam}.$$

Now we can formulate the main theorem of this section—a theorem of functorness of $\mathfrak{A}$-equlibrium. Let $(\Gamma, \mathfrak{A}) \in \mathrm{Ob}\Delta E(C)$, $A \in C$. Denote by $\mathrm{Eq}(A, \Gamma, \mathfrak{A})$ the set of $\mathfrak{A}$-equilibrium points in the A-realization of the game Γ.

Theorem 3.9 Eq *is a (contravariant in both variables) functor from the category* $C \times \Delta E(C)$ *to* **Set**[1].

Proof. Let $\Gamma = \langle I, \{S_i\}, \{K_i\}, \{\varphi_i\}_{i\in I}\rangle$ be a game from the category $\Delta(C)$, $\mathfrak{A}$ be a behaviourial type in it and $s^* \in \mathrm{Hom}(A, S)$ be an $\mathfrak{A}$-equilibrium point in $(\Gamma, \mathfrak{A})$, i.e. $\varphi_i s^* \geq \varphi_i \bar{\alpha}_i s^*$ for every $i \in I$ and every $\alpha_i \in \mathfrak{A}_i$.

Let $g : B \to A$, then the morphism $\Phi_C(g, \Gamma) = (1, \{f_i\}, \{f^i\}_{i\in I})$ sends the game $\Gamma(B)$ to the game $\Gamma(A)$. We shall prove that fs^* is an $\mathfrak{A}$-equilibrium point in $\Gamma(B)$. Since h_{K_i} is a functor to the category **OrdSet**, the equilibrity of s^* implies that

$$h_{K_i}(g)(\varphi_i s^*) \geq h_{K_i}(g)(\varphi_i \bar{\alpha}_i s^*),$$

but

$$\begin{aligned} h_{K_i}(g)(\varphi_i s^*) &= \varphi_i f s^*, \\ h_{K_i}(g)(\varphi_i \bar{\alpha}_i s^*) &= \varphi_i \bar{\alpha}_i f s^*, \end{aligned}$$

therefore for every α_i we have $\varphi_i f s^* \geq \varphi_i \bar{\alpha}_i f s^*$, and fs^* is an $\mathfrak{A}$-equilibrium point.

Now let $(F, \{f_i\}, \{f^i\}_{i\in I}) \in \mathrm{Hom}_{\Delta E(C)}(\Gamma, \Gamma')$. It suffices to prove that if t^* is a $\mathfrak{A}'$-equlibrium point in A-realization of Γ' (where A is an arbitrary object of C), then ft^* is an $\mathfrak{A}$-equilibrium point in A-realization of Γ. Let $\alpha_i \in \mathfrak{A}_i$. Consider the outcome $ft^*||\alpha_i(ft^*)_i$ (here $(ft^*)_i$ is the i-th component of outcome ft^*). By definition of compatibility of the behaviorial types $\alpha_i(ft^*)_i \in f^*_{(F(i))}(\mathfrak{A}'_{F(i)})$, i.e. there exists $\beta_{F(i)} \in \mathfrak{A}'_{F(i)}$ such that $\alpha_i(ft^*)_i = f_i \beta_{F(i)}(t^*)_{F(i)}$ and $f_{i_1}\beta_{F(i)} = f_{i_1}$ for all other i_1 from $F^{-1}(F(j))$. Therefore

$$f\,t^*||\alpha_i(f\,t^*)_i = f(t^*||\beta_{F(i)}(t^*)_{F(i)}).$$

Since t^* is an $\mathfrak{A}'$-equilibrium point in $\Gamma(A)$, we have

$$\psi_{F(i)}(t^*) \geq \psi_{F(i)}(t^*||\beta_{F(i)}(t)_{F(i)}).$$

Monotonicity of f^i implies

$$f^i\psi_{F(i)}(t^*) \geq f^*\psi_{F(i)}(t^*||\beta_{F(i)}(t^*)_{F(i)}).$$

Finally, by commutativity of the corresponding diagram we have

$$\varphi_i f t^* \geq \varphi_i(ft^*||\alpha_i(ft^*)_i),$$

and it means that the A-outcome ft^* is an $\mathfrak{A}$-equilibrium point in the game Γ. □

Corollary 3.10 Eq *is a contravariant functor from* **Gam** *to* **Set**.

[1] For some generalizations of this theorem see: Lapitsky V. On some categories of games and corresponding equilibria. Int. Game Theory Rev.,1999, vol. 1, N 2, pp. 169–185.

4 Optimality principles in the category Gam

In this section we shall consider the games in the category **Gam**. Our purpose is to give a general definition of an optimality principle for such games. First of all, every optimality principle F has to be a mapping F : Ob **Gam** $\to$ Ob**Set**. The first natural restriction we want to impose on F is its functorness. In our context it may be considered as a reflection of the general character—of *universality*—of an optimality principle. More precisely, let $\mathbf{S}$ be the natural contravariant functor **Gam** $\rightsquigarrow$ **Set** sending a game Γ to its set of outcomes S. Then the first axiom is:

A1. F is a subfunctor of $\mathbf{S}$.

The second axiom ought to reflect the optimalness of optimality principle. Let $\mathbf{G(1)}$ be the full subcategory of **Gam** formed by all games with only one player (i.e. by optimization problems) and $(\Gamma_0, \mathfrak{A}_0) \in \text{Ob } \mathbf{G(1)}$, $\Gamma_0 = \langle S, K, \varphi \rangle$; $\mathfrak{A}_0(S) = \{s' \in S / \exists \alpha_0 \in \mathfrak{A}_0 \; s' = \alpha_0 s\}$. Let $E(s)$ denote the set of outcomes $s' \in \mathfrak{A}_0(s)$ such that $\varphi(s')$ is a maximum element of the set $\varphi(\mathfrak{A}_0(s))$, and $E(\Gamma_0, \mathfrak{A}_0) = \cup_{s \in S} E(s)$. It is clear that the restriction of the functor Eq on the category $\mathbf{G(1)}$ coincides with E. Now we can formulate our second axiom.

A2. On the category $\mathbf{G(1)}$ F coincides with E.

This axiom postulates a kind of "individual rationality", the weakest rationality in the game theory: in the case of no opponents the only player seeks to maximize his payoff (and not, for example, to minimize it).

Now we can give our basic definition.

Definition 4.1 A map F : Ob **Gam** $\to$ Ob **Set** is said to be an optimality principle iff it satisfies both axioms **A1** and **A2**.

Corollary 3.10 means that Eq is an optimality principle. Naturally it is not unique, and we shall give some other examples.

We shall say that an optimality principle F is *stronger* than an optimality principle G if the inclusion $F(\Gamma, \mathfrak{A}) \subset G(\Gamma, \mathfrak{A})$ holds for every game $(\Gamma, \mathfrak{A})$ from **Gam**. Now we want to construct the strongest and the weakest optimality principles in the category **Gam**.

Let $s = (s_1, \dots, s_n)$ be an outcome of the game $(\Gamma, \mathfrak{A})$. It is said to be a local maximum if for every $\alpha_1, \dots, \alpha_n$, $\alpha_i \in \mathfrak{A}_i$, and every $i \in I$

$$\varphi_i(\alpha_1(s), \dots, \alpha_n(s_n)) \leq \varphi_i(s).$$

Theorem 4.2 *Let* $\mathrm{L}(\Gamma, \mathfrak{A})$ *be the set of local maxima of the game* $(\Gamma, \mathfrak{A})$. *Then* L *is the strongest optimality principle in the category* **Gam**.

Proof. First of all we have to prove that L is an optimality principle. Clearly, L satisfies **A2**, and it suffices to show that L is a functor. Let $(\mathrm{P}, f_i, f^i) : (\Gamma, \mathfrak{A}) \to (\Gamma', \mathfrak{A}')$, and $t \in \mathrm{L}(\Gamma', \mathfrak{A}')$; then we have to show that $f(t) \in \mathrm{L}(\Gamma, \mathfrak{A})$.

For an arbitrary collection $\alpha_1, \ldots, \alpha_n$, $\alpha_i \in \mathfrak{A}_i$, and $i \in I$, consider an element of the set K_i

$$\varphi_i(\alpha_1(f(t))_1, \alpha_2(f(t))_2, \ldots, \alpha_n(f(t))_n).$$

By definition of morphism of games, for every $j \in \mathrm{P}(I)$ there exist $\beta_j \in \mathfrak{A}'_j$ such that $j = \mathrm{P}(i)$ and $\alpha_i f_i = f_i \beta_j$. Then

$$\begin{aligned} &\varphi_i(\alpha_1(f(t))_1, \alpha_2(f(t))_2, \ldots, \alpha_n(f(t))_n) = \\ &\quad = \varphi_i(\alpha_1 f_1(t_{\mathrm{P}(1)}), \ldots, \alpha_n f_n(t_{\mathrm{P}(n)})) = \\ &= \varphi_i(f_1 \beta_{F(1)}(t_{\mathrm{P}(1)}), \ldots, f_n \beta_{\mathrm{P}(n)}(t_{\mathrm{P}(n)})) = \\ &\quad = \varphi_i(\alpha_1 f_1(t_{\mathrm{P}(1)}), \ldots, \alpha_n f_n(t_{\mathrm{P}(n)})) = \\ &\qquad = \varphi_i f(\beta_1(t_1), \ldots, \beta_{n'}(t_{n'})). \end{aligned}$$

Furthermore, since $t \in \mathrm{L}(\Gamma', \mathfrak{A}')$

$$\psi_{\mathrm{P}(i)}(\beta_1(t_1), \ldots, \beta_{n'}(t_{n'})) \leq \psi_{\mathrm{P}(i)}(t)$$

and

$$f^i \psi_{\mathrm{P}(i)}(\beta_1(t_1), \ldots, \beta_{n'}(t_{n'})) \leq f^i \psi_{\mathrm{P}(i)}(t).$$

By definition of morphism, $f^i \psi_{\mathrm{P}(i)} = \varphi_i f$, and so we have

$$\varphi_i f(\beta_1(t_1), \ldots, \beta_{n'}(t_{n'})) \leq \varphi_i f(t),$$

that is

$$\varphi_i(\alpha_1(f(t))_1, \alpha_2(f(t))_2, \ldots, \alpha_n(f(t))_n) \leq \varphi_i f(t)$$

and $f(t) \in \mathrm{L}(\Gamma, \mathfrak{A})$; the functorness of L is proved.

To demonstrate its minimality, we shall construct for every game Γ and a local maximum s in it a one-person game Γ^1 and a morphism $\Gamma \to \Gamma^1$ such that $f(s^1) = s$ for some local maximum s^1 in Γ^1.

We define this game $\Gamma^1 = \langle T, \leq, \mathfrak{A}^1 \rangle$ as follows:

$$T = S;$$

$$\mathfrak{A}^1 = \{\alpha \in \mathrm{Hom}(S, S) / \alpha(s) = (\alpha_1(s_1), \ldots, \alpha_n(s_n)), \alpha_i \in \mathfrak{A}_i\};$$

$$s' \leq s'' \text{ iff } s = s'' \text{ and there is } \alpha \in \mathfrak{A}^1 \text{such that } s' = \alpha(s).$$

It is clear that $s^1 = s$ is a local maximum in this game. Now we define a morphism

$(\mathrm{P}, f_i, f^i) : \Gamma \to \Gamma^1$.

P is a projection of I onto one point;

$f_i : t \to S_i$ is a projection on i-th component (so f is identical);

$f^i = \varphi_i$ (hence, it is monotone).

It is easy to see that all conditions required in the definition of a morphism of the games are fulfilled and this is really a morphism of category **Gam**. Therefore, since $f(s^1) = s$, our theorem is proved. □

Let once more $s = (s_1, \ldots, s_n)$ be an outcome of the game $(\Gamma, \mathfrak{A})$. We shall denote the elements of the set $\bar{S}_i = \Pi_{j \neq i} S_j$ by $\bar{s}_i$; an outcome s is then a pair $(\bar{s}_i, s_i)$. We define the (binary) relation on the set S_i by

$$s'_i \prec s_i \Longleftrightarrow \exists \bar{s}'_i, \bar{s}_i \quad \varphi_i(\bar{s}'_i, s'_i) \leq \varphi_i(\bar{s}_i, s_i).$$

We shall say that a strategy s'_i is *potentially more dangerous* than a strategy s_i if there exists a sequence $s_{i_0} \prec s_{i_1} \prec \cdots \prec s_{i_m}$ such that $s_{i_0} = s'_i$, $s_{i_m} = s_i$.

Let

$$S_i^0 = \{s_i \in S_i | \forall\ \alpha_i \in \mathfrak{A}_i\ \alpha_i(s_i) \text{ is potentially more dangerous than } \mathrm{s_i}\}$$

and $\mathrm{l}(\Gamma) = \Pi_\mathrm{i} \mathrm{S}_\mathrm{i}^0$.

Then we have

Theorem 4.3 $\mathrm{l}(\Gamma, \mathfrak{A})$ *is the weakest optimality principle in the category* **Gam**.

The proof of this theorem is more or less similar to the proofs of the previous theorem and Theorem 5.2 below, and we omit it.

We obtain therefore a triple of optimality principles in the category **Gam** : $\mathrm{L} \subset \mathrm{Eq} \subset \mathrm{l}$. Another series of optimality principles is defined by the following observation (which holds for a general basic category C).

Let $\Gamma \in \mathrm{Ob}\Delta E(e)$, $K \subset I$ be a coalition (subset of players), α is an arbitrary element of $\Pi_{i \in K} \mathfrak{A}_i \subset \Pi_{i \in K} \mathrm{Hom}(S_i, S_i) \subset \mathrm{Hom}(\Pi_{i \in k} S_i, \Pi_{i \in K} S_i)$, and $\bar{\alpha}_k$ is an induced morphism $S \to S$. We shall suppose that once formed coalition K can transform an outcome $s : A \to S$ into an outcome $\bar{\alpha}_k S$. We define a series of cooperative equilibria: A-outcome s of the game $(\Gamma, \mathfrak{A})$ is said to be a *k-equilibrium point* (where k is a positive integer) iff for every $K \subset I$ such that $|K| = k$ and every α_k we have $\varphi_i \bar{\alpha}_k s \leq \varphi_i s\ \forall i \in K$. We denote by $\mathrm{Eq}^k(\Gamma, \mathfrak{A}, A)$ the set of all k-equilibrium A-points of Γ. It is clear that the family $\mathrm{Eq}^k(\Gamma, \mathfrak{A}, A)$ forms a kind of filtration from classical Nash equilibrium to the strong equilibrium. Now, we have

Theorem 4.4 Eq^k *is an optimality principle in the category* **Gam** *for all* k.

Taking into consideration that the cardinality of a set does not increase under any mapping, the proof of this theorem is similar to the proof of Theorem 3.9.

5 Axiomatics of equilibrium

Thus, equilibrium is only one of the series of optimality principles in the category **Gam**. Now we would like to introduce some additional axioms that would allow us to select it (or its modifications) from this series. For this purpose we define a reduction of the game Γ – the game $R(\Gamma, \mathfrak{A})$; all its components are the same as in $(\Gamma, \mathfrak{A})$ except for the sets K_i or, more precisely, the corresponding orders on S. We shall define them as follows:

$$x \underset{i}{\geq} y \Longleftrightarrow \exists s_i \in S_i \quad x = y || s_i \quad \varphi_i(x) \geq \varphi_i(y).$$

Now we can state a complementary axiom.

A3. $F = FR$.

Clearly, this axiom is not as natural as the previous ones; it postulates that an optimality principle F is defined only by the preferences between outcomes which are interchangeable for the corresponding player. As every player considers in this case the preferences only between outcomes he can transform one into another and doesn't take into account the possibilities of other players, this axiom reflects the noncooperative character of F, eliminating any kind of cooperation.

Now we have the following theorem:

Theorem 5.1 Eq *is the strongest optimality principle satisfying* **A3**.

Proof. First of all, it is clear that $\mathrm{Eq} = \mathrm{Eq}R$. Now, let F be a functor satisfying axiom **A3**. It is sufficient to show that $s \in \mathrm{Eq}(\Gamma)$ implies $s \in F(\Gamma)$ for any game $(\Gamma, \mathfrak{A}) \in \mathrm{Ob}\mathbf{Gam}$.

For this purpose we construct some auxiliary games of category **Gam**:

1) $R(\Gamma)$;

2) Γ^1 – is a one-person game:
$\Gamma^1 = \langle \{a\}, S^1, \leq, \mathfrak{A}^1 \rangle$,
$S^1 = S$,
$\mathfrak{A}^1 = \{\alpha/\bar{\alpha}_1\bar{\alpha}_2 \ldots \bar{\alpha}_k,\ \alpha_j \in \mathfrak{A}_j,\ j = 1, \ldots, n\}$,
$s^1 \geq s^2 \Longleftrightarrow s^1 = s$ and $\exists \alpha \in \mathfrak{A}^1\ s^2 = \alpha(s^1)$;

3) $\Gamma' = \langle I, \{S_i\}, \{\mathfrak{A}_i\}, \{\underset{i}{\leq'}\}_{i\in I}\rangle$,

$\underset{i}{\leq'}$ is a combination of two orders on the set S: the order $\underset{i}{\leq}$ of the game $R(\Gamma)$ and the order $\leq$ of the game Γ^1;

4) $R(\Gamma')$.

For any game Γ there exists a canonical morphism of the category **Gam** $\Gamma \to R(\Gamma)$ (if the game is defined in the form of the preferences orders of players, all its components are identical); we denote it by $r(\Gamma)$. It is clear that in this case we have $F(r(\Gamma)) = 1_{F(\Gamma)}$.

To prove the statement of the theorem we consider the following diagram:

$$\begin{array}{ccc} \Gamma & \stackrel{r(\Gamma)}{\longrightarrow} & R(\Gamma) \\ & & \downarrow \\ \Gamma' & \stackrel{r(\Gamma')}{\longrightarrow} & R(\Gamma') \\ \downarrow & & \\ \Gamma^1 & & \end{array}$$

First of all, it is clear that $R(\Gamma) = R(\Gamma')$. In fact, the functor F "annihilates" all complementary (as for $R(\Gamma)$) ordered pairs in the game Γ', since they differ by strategy of at least one player except for the one under consideration. Therefore, let the right vertical arrow of the diagram be $1_{R(\Gamma)}$. The functor F transforms the diagram into a morphism of games $F(H) : F(\Gamma^1) \to F(\Gamma)$. Since $s \in F(\Gamma^1)$ by construction of Γ^1, it remains to construct a morphism of games $(H, h_i, h^i) : \Gamma' \to \Gamma^1$ such that $h(s) = s$. This morphism is defined as follows:

$H : I \to \{a\}$,

$h_i : S \to S_i$ is a projection on the i-th component,

$h^i : S \to S$ is identical.

We complete the proof by observing that this collection is indeed a morphism of the category **Gam** (i.e. it is consistent with the behaviorial types). □

Now we shall construct the weakest optimality principles in the category **Gam** satisfying the axiom **A3**.

We call an outcome s of the game $\Gamma \in \mathrm{Ob}\mathbf{Gam}$ *bearable for a player* i if the following condition holds:

$$\varphi_i(s||\alpha_i(s_i)) \leq \varphi_i(s) \Longrightarrow \exists s_{\hat{i}}\ \varphi_i(\bar{\alpha}_i(s)||s_{\hat{i}}) \leq \varphi_i(s||s_{\hat{i}}).$$

An outcome s is called *bearable* if it is bearable for every player i. We denote the set of all bearable outcomes of the game Γ by $\mathrm{n}(\Gamma)$. Then we have

Theorem 5.2 n *is the weakest optimality principle satisfying* **A3**.

Proof. First of all, n satisfies the axiom **A3** because of only "i-reducible" outcomes are compared in the definition of a bearable outcome.

Next, we want to demonstrate that n is an optimality principle. Let (F, f_i, f^i) be a morphism $(\Gamma, \mathfrak{A}) \to (\Gamma', \mathfrak{A}')$ and $t \in \mathrm{n}(\Gamma', \mathfrak{A}')$. We have to show that $s = f(t) \in \mathrm{n}(\Gamma, \mathfrak{A})$. Let there exist $\alpha_i \in \mathfrak{A}_i$ such that $\varphi_i(s\|\alpha_i(s_i)) \not\leq \varphi_i(s)$. As usual we select $\beta_{F(i)} \in \mathfrak{A}'_{F(i)}$ such that $\alpha_i f_i = f_i \beta_{F(i)}$ and $f_j \beta_{F(i)} = f_j$ for any other j from $F^{-1}(F(i))$.

Then we have

$$\psi_{F(i)}(t\|\beta_{F(i)}(t_{F(i)})) \leq \psi_{F(i)}(t),$$

and there exists a family $t'_1, \dots, t'_{n'}$; $t'_j \in T_j$ such that

$$\psi_{F(i)}(t'_1, \dots, \beta_{F(i)}(t_{F(i)}), \dots, t'_{n'}) \leq \psi_{F(i)}(t'_1, \dots, t_{F(i)}, \dots, t'_{n'}).$$

Hence, by definition of morphism of category **Gam** we have

$$\varphi_i(s'_1, \dots, \alpha_i(s_i), \dots, s'_n) \leq \varphi_i(s'_1, \dots, s_i, \dots, s'_n)$$

where $s'_k = f_k(t'_{F(k)})$, i.e. $s \in \mathrm{n}(\Gamma)$.

Now, we prove the maximality of n. Let s be an outcome in the game Γ (considered in the form of preferences on the set S) and $s \notin \mathrm{n}(\Gamma, \mathfrak{A})$. Then the axiom **A3** implies that $s \notin \mathrm{n}(\mathrm{R}(\Gamma))$. Thus it is sufficient to construct a one-person game $(\Gamma^1, \mathfrak{A}^1)$ and a morphism $(F, f_1, f^1) : \Gamma^1 \to \mathrm{R}(\Gamma)$ such that $s^1 = f(s) \notin \mathrm{E}(\Gamma^1, \mathfrak{A}^1)$. Let s be not-bearable for a player $i \in I$, that is for some $\alpha_i \in \mathfrak{A}_i$ $\varphi_i(s\|\alpha_i(s_i)) \leq \varphi_i(s)$, and there are no $s'_1, \dots, s'_n$ ($s_j \in S_j$, $j \neq i$) such that

$$\varphi_i(s'_1, \dots, \alpha_i(s_i), \dots, s'_n) \leq \varphi_i(s'_1, \dots, s_i, \dots, s'_n).$$

Then we define $\Gamma^1 = \langle \{a\}, S_i, \mathfrak{A}_i, \leq \rangle$ where the preference $\leq$ on the set S_i is the image of the preference $\leq_i$ on S under canonical projection $pr_i : S \to S_i$. Then, by choice of s and α_i

$$pr_i(s) \not\geq pr_i(s\|\alpha_i(s_i)),$$

i.e. in the game Γ' an outcome $s^1 = pr_i(s)$ does not belong to the set $\mathrm{E}(\Gamma^1, \mathfrak{A}^1)$.

Now, we define morphism (F, f_1, f^1) by

$$F : \{a\} \to i$$
$$, \quad f_1 : S_i \xrightarrow{\sim} S_i$$
$$f^1 : S \xrightarrow{pr_i} S_i$$

It is clear, that $f(s) = s^1$, q.e.d. □

Remark With the corresponding modification of **A3** Theorem 5.1 and Theorem 5.2 can be generalized to the case of cooperative equilibria Eq^k.

Acknowledgement This chapter is the author's translation of V. Lapitsky, K teorii printsipov optimal'nosty v beskoalitsionnyh igrah. Mathematical Methods in Social Sciences. Vilnius, vol. 18, 35-56. 1985.

References

1. Gelfand S., and Yu. Manin (1997): *Methods of Homological Algebra.* Berlin: Springer.

2. Grothendieck, A. (1957): "Sur quelques points d'Algebre homologique," *Tohoku Math. J.*, 9, 119–221.

3. Lapitsky, V. (1983): "On some classes of morphisms of noncooperative games," *Multistage, differential, noncooperative and cooperative Games.* Kalinin, 55–61 (in Russian).

4. MacLane, S. (1971): *Categories for the Working Mathematician.* Berlin: Springer.

5. Takahara Y., B. Nakano, and K. Kijima (1980): "A unified theory of decision principles," *Int. J. Systems Sci.*, 11, 1295–1314.

6. Takahara Y., B. Nakano, and K. Kijima (1981): "A structure of rational decision principles," *Int. J. General Systems*, 7, p. 175–188.

7. Vorob'ev, N. (1994): *Foundations of Game Theory: Noncooperative Games.* Basel: Birkhäuser.

Part II

Cooperative game theory

Chapter 5

A sufficient condition for the coincidence of the core of a cooperative game with its solution

G. Diubin

Abstract: A sufficient condition of the coincidence of the core and the solution (von Neumann-Morgenstern stable set) of a cooperative game is given. This condition is formulated as a set of upper bounds for the characteristic function which cannot be strengthened within the class of symmetric games.

Key words: Core, stable set.

We consider an n-person cooperative game determined by its characteristic function v in 0-1-reduced form. Superadditivity of v is not imposed. It was shown by Gillies (1959) that if $v(S) \leq \frac{1}{n}$ for all $S \neq N$ (where N is the grand coalition) then the core is the (unique) stable set (in the sense of von Neumann and Morgenstern). Bondareva (see also Bondareva, 1963) informed the author that this assertion holds if $v(S) \leq \frac{1}{n-1}$ for all coalitions $S \neq N$. In this note we prove the following theorem.

Theorem 1 *Let (N, v) be a cooperative n-person game in 0-1-reduced form. Let for any coalition S*

$$v(S) \leq \frac{1}{n - |S| + 1}$$

where $|S|$ is the number of players of the coalition S. Then the core of the game (N,v) coincides with its (unique) stable set.

Proof. Consider an arbitrary imputation x not belonging to the core. Let S_0 be a coalition with a minimal number of players such that $v(S_0) > \sum_{i\in S_0} x_i$. Let k be the cardinality of the coalition S_0. Obviously, $2 \le k \le n-1$. We construct the imputation y as follows:

$$y_i = \begin{cases} x_i + \frac{v(S_0)-\sum_{i\in S_0} x_i}{k} & i \in S_0 \\ \frac{1-v(S_0)}{n-k} & i \notin S_0. \end{cases}$$

It can be easily seen that $\sum_{i\in S_0} y_i = v(S_0)$ and $y_i > x_i$, $i \in S_0$. Therefore y dominates x over the coalition S_0. We now prove that y is in the core. For this, it suffices to show that $\sum_{i\in S} y_i \ge v(S)$ for an arbitrary S. Let firstly $|S| \le k$. Note that y is not dominated over any subset S' of S_0 because then $\sum_{i\in S'} y_i \ge v(S')$. If at least one player from S is not in S_0 then

$$\sum_{i\in S} y_i \ge \frac{1-v(S_0)}{n-k} \ge \frac{1}{n-k+1} \ge \frac{1}{n-|S|+1} \ge v(S).$$

Thus, y is not dominated over any coalition S containing at most k players.

Let now $|S| > k$. If $S \supset S_0$ then

$$\begin{aligned} \sum_{i\in S} y_i &\ge \frac{((S)-k)(1-v(S_0))}{n-k} + v(S_0) \ge \\ &\ge \frac{|S|-k}{n-k} \ge \frac{1}{n-|S|+1} \ge v(S). \end{aligned}$$

If S does not contain S_0 then the number of players of S not in S_0 is at least $|S|-k+1$, and therefore

$$\begin{aligned} \sum_{i\in S} y_i &\ge \frac{(|S|-k+1)(1-v(S_0))}{n-k} \ge \\ &\ge \frac{|S|-k+1}{n-k+1} \ge \frac{1}{n-|S|+1} \ge v(S). \end{aligned}$$

Thus, we have shown that y is not dominated over any coalition S. Hence, y is in the core. Since y dominates x, the theorem is proved. □

Let (N,v) be a symmetric game with

$$v(S) = \begin{cases} r > \frac{1}{n-m+1} & |S| = m \\ 0 & |S| \ne m, \end{cases}$$

for some $1 < m < n$ and some $r \in \mathbb{R}$. Then, evidently, the imputation

$$x = \left(\frac{1}{n-m+1}, \dots, \frac{1}{n-m+1}, 0, \dots, 0\right)$$

which is not in the core, cannot be dominated by any imputation from the core. This example shows within the class of symmetric games the inequalities stated in the theorem cannot be strenghtened to ensure the validity of the theorem.

Acknowledgement This chapter is a translation of the orginal article, which appeared in: Teoria igr, Proceedings of the first All-union conference on game theory, Erevan, Press of Academy of Sciences of Armenian SSR, 1973, 154-155.

References

1. Bondareva, O.N. (1963): “Some applications of linear programming methods to cooperative game theory,” *Problemy Kibernetiki*, 10 (in Russian).

2. Gillies, D.B. (1959): “Solutions to general non-zero-sum games,” *Contributions to the theory of games*, II, Ann. Math. Studies 40.

Chapter 6

On the Shapley function for games with an infinite number of players

G. Diubin

Abstract: A well-known theorem due to Keane (1969) states that for a finite cooperative game, the Shapley vector is the solution of a minimization problem for a certain quadratic functional. For cooperative games with a countable number of players we introduce an analogue of Keane's functional and define the Shapley value as the solution of the corresponding minimization problem. It is shown that this definition is consistent, and the solution is unique.

Key words: Shapley value, infinitely many players.

In Aumann and Shapley (1976) an axiomatic definition of the Shapley function for games with an infinite number of players is given, and several theorems concerning the existence of this function are proved. In Keane (1969) it is shown that the Shapley value for a game with a finite number of players can be defined as a measure on a finite set such that the distance (in a certain metric) to the characteristic function is minimal. In this paper, similarly to Keane (1969), we give a definition of the Shapley value for games with an infinite number of players; the Shapley function is defined as a function whose distance (in a metric generalizing that from Keane, 1969) to the characteristic function is minimal. We assume that the characteristic function v is monotone, that is, if $S_1 \supseteq S_2$ then $v(S_1) \geq v(S_2)$ for all $S_1, S_2 \subseteq N$.

We introduce the necessary notations. Let J be the set of players, and 2^J the direct product of two-element sets $\{0,1\}$ over the index set J. Then to each subset T of J there corresponds a point $S \in 2^J$ whose coordinates for all players from this subset are equal to 1, and for the remaining players they are zeros. The converse is also evident: to each point $S \in 2^J$ there corresponds a subset $T \subset J$. In the sequel we shall make no distinction between the points of the set 2^J and coalitions.

The function v defined on the set 2^J is called a characteristic function. It is supposed that the game is given in 0-reduced form, i.e. the function v equals zero on all one-element coalitions. Any non-negative countably additive measure x on the set of players with a countable carrier satisfying $x(J) = v(J)$ is called an imputation (the σ-algebra is the set of all subsets of J). Further we shall suppose for simplicity that the set J is countable. If we suppose that the carrier of any imputation is countable then all proofs will be valid also for an arbitrary J. We denote the set of all imputations by X. This set coincides with the set of all functions on 2^J satisfying the conditions:

a) $x(S) \geq 0$ if S is a one-element coalition;

b) $x(S) = v(J)$ if S is the coalition of all players;

c) $x(\sum_i S_i) = \sum_i x(S_i)$ if the coalitions S_i are disjoint. The sum of two points from 2^J is their coordinate-wise logical sum (to every imputation x there corresponds the function $x(S) = \sum_{i \in S} x_i$, $x_i = x(\{i\})$).

We define a measure μ on the set 2^J as follows. We let that the measure of any cylinder $\mathfrak{S}_n$ in which n coordinates are fixed and the remaining ones are arbitrary, equals

$$\frac{(s-1)!(n-s-1)!}{(n-1)!}$$

where s is the number of ones among the fixed coordinates, $s \neq 0$, $s \neq n$; otherwise we suppose that the measure of this cylinder in infinite. The measure defined in this way is consistent on these cylinders. For the proof, it is sufficient to check that

$$\mu(\mathfrak{S}_{n-1}) = \mu(\mathfrak{S}_n^0) + \mu(\mathfrak{S}_n^1)$$

where $\mathfrak{S}_{n-1}$ is an arbitrary cylinder with $n-1$ coordinates fixed, $\mathfrak{S}_n^0$ a cylinder with n coordinates fixed where $n-1$ coordinates coincide with the fixed coordinates of the cylinder $\mathfrak{S}_{n-1}$, and the n-th fixed coordinate is 0; $\mathfrak{S}_n^1$ an analogous cylinder with the n-th fixed coordinate 1. Indeed, if s is the number of ones among the fixed coordinates of the cylinder $\mathfrak{S}_{n-1}$ and $s \neq 0$, $s \neq n-1$, then

$$\begin{aligned}
\mu(\mathfrak{S}_{n-1}) &= \frac{(s-1)!(n-s-2)!}{(n-2)!}, \\
\mu(\mathfrak{S}_n^0) &= \frac{(s-1)!(n-s-1)!}{(n-1)!}, \\
\mu(\mathfrak{S}_n^1) &= \frac{s!(n-s-2)!}{(n-1)!},
\end{aligned}$$

and the desired equality can be verified by a simple calculation. If $s=0$ or $s=n-1$ then the measure of the cylinder $\mathfrak{S}_{n-1}$ is infinite. But in this case, one of the cylinders $\mathfrak{S}_n^0$ or $\mathfrak{S}_n^1$ in the right-hand side of the equality to be proved also has an infinite measure. It can be shown that this measure on cylinders determines a unique measure μ on the σ-algebra generated by these cylinders over the measure μ. Evidently, the function $x(S)$ is continuous on the space 2^J equipped with the Tychonoff topology. If the set of players is at most countable, then the continuity of the function $x(S)$ implies its measurability (in this case there is a countable base of open sets; hence, the imputation for any set of players will be measurable, because each set is determined by a countable number of coordinates).

Let N be some finite subset of the set of all players. Denote by $\mathfrak{S}_N^0$ a cylinder in which all coordinates of players from N are zeros, and the remaining ones are arbitrary. Similarly, $\mathfrak{S}_N^1$ is a cylinder in which the coordinates of players from N equal 1, and the remaining ones are arbitrary. Let φ be a mapping of 2^J which assigns to each point $S=(l_1,\ldots,l_n,\ldots)$ the point $(1-l_1,\ldots,1-l_n,\ldots)$ where $l_i \in \{0,1\}$.

Lemma 1 *The mapping φ is measure preserving. It transforms the cylinders of the form $\mathfrak{S}_N^0$ into cylinders of the form $\mathfrak{S}_N$.*

The proof of the first assertion in this lemma follows immediately from the fact that the measure of a cylinder with n fixed coordinates, from which exactly s coordinates are equal to 1, equals the measure of a cylinder having $n-s$ ones among the n fixed coordinates. The second assertion of the lemma is obvious.

In the sequel, the following result is used.

Theorem 2 (Keane, 1969) *For a finite game determined by the characteristic function v, the minimum of the function*

$$\sum_{\substack{S \subset N \\ s \neq 0 \\ s \neq |N| = n}} (s-1)!(n-s-1)!(v(S)-x(S))^2$$

on the hyperplane $\sum_i x_i = v(N)$ is attained on the Shapley vector.

If the game is given in 0-reduced form and the function v is monotone then the Shapley vector is an imputation.

Similarly to this, we can define the Shapley function (which will be an imputation) as follows.

Definition 3 If the minimum of the functional

$$\int_{2^J} (v(S) - x(S))^2 d\mu(S) \tag{1}$$

on the set X exists, then the imputation on which this minimum is attained is called the Shapley imputation.

Below we shall show that this definition is quite natural.

Let N be some finite subset of the set of all players on which the characteristic function v is given. We define the characteristic function v_{imm} on the set of all players as $v_{\text{imm}}(S) = v(S \cap N)$. We call the game given by this characteristic function the immersion of a finite game into the infinite one. It is natural to suppose that the Shapley imputation of this game is a measure on the set N and coincides with the Shapley vector of the finite game determined by the characteristic function v.

Theorem 4 *The Shapley imputation of the immersion (J, v_{imm}) of a finite monotone game (N, v) exists and coincides with the Shapley vector of the latter on the set N. The players not in N get nothing.*

Proof. Let $X(v_{\text{imm}})$ be the set of elements from $L_2(2^J, \mu)$ of the form $v_{\text{imm}}(S) - x(S)$, and let $X_f(v_{\text{imm}})$ be the subset of $X(v_{\text{imm}})$ defined by the imputations on a finite set of players. We show that $X_f(v_{\text{imm}})$ is dense in $X(v_{\text{imm}})$. If $v_{\text{imm}} - x \in L_2(2^J, \mu)$ then

$$\int_{\mathfrak{S}_N^0} (v_{\text{imm}}(S) - x(S))^2 d\mu(S) < \infty.$$

We have $v_{\mathrm{imm}} \equiv 0$ on $\mathfrak{S}_N^0$. Hence

$$\int_{\mathfrak{S}_N^0} x^2(S) d\mu(S) < \infty.$$

This means that for any ε we can find a finite set of players $N_1 \supseteq N$ such that

$$\int_{\mathfrak{S}_{N_1}^0} x^2(S) d\mu(S) < \varepsilon \tag{2}$$

(we take into account that the imputation x is concentrated on a countable set of players).

Let i_0 be some player from the set N. For a finite set M, $J \supseteq M \supseteq N_1$ we define the imputation x^M as

$$x_i^M = \begin{cases} x_i + v(J) - x(M), & i = i_0 \\ x_i, & i \in M \setminus \{i_0\} \\ 0, & i \notin M. \end{cases}$$

Then, on the set $2^J \setminus (\mathfrak{S}_{N_1}^0 \cup \mathfrak{S}_{N_1}^1)$ which has a finite measure, the function x^M converges uniformly to the function x if $v(J) - x(M)$ tends to zero when $|M| \to \infty$. The latter condition can be easily satisfied, since the measure x is concentrated on an at most countable set of players. Therefore we can find a finite set M, which is sufficiently "powerful", such that

$$\int_{2^J \setminus (\mathfrak{S}_{N_1}^0 \cup \mathfrak{S}_{N_1}^1)} (x(S) - x^M(S))^2 d\mu(S) < \varepsilon. \tag{3}$$

Taking into account that

$$x(\varphi(S)) = v(J) - x(S)$$

for any imputation X, we get

$$(x(S) - x^M(S))^2 = (x(\varphi(S)) - x^M(\varphi(S)))^2.$$

This equality together with Lemma 1 implies that

$$\int_{\mathfrak{S}_{N_1}^0} (x(S) - x^M(S))^2 d\mu(S) = \int_{\mathfrak{S}_{N_1}^1} (x(S) - x^M(S))^2 d\mu(S). \tag{4}$$

On the other hand, equality (2) implies

$$\int_{\mathfrak{S}_{N_1}^0} (x(S) - x^M(S))^2 d\mu(S) \leq \int_{\mathfrak{S}_{N_1}^0} x^2(S) d\mu(S). \tag{5}$$

Taking into account (3), (4) and (5), we get

$$\int_{2^J}[(v_{\text{imm}}(S) - x(S)) - (v_{\text{imm}}(S) - x^M(S)]^2 d\mu(S) = \\ = \int_{2^J}(x(S) - x^M(S))^2 d\mu(S) \leq 3\varepsilon.$$

Since we can take an arbitrarily small ε, this proves the density of $X_f(v_{\text{imm}})$ in $X(v_{\text{imm}})$. Furthermore, the set $X_f(v_{\text{imm}})$ (and hence the set $X(v_{\text{imm}})$ is nonempty. Indeed, if x is an imputation on some set $M \subset N$ then

$$\int_{2^J}(v^M_{\text{imm}}(S) - x(S))^2 d\mu(S) = \\ = \sum_{\substack{S \subset M \\ s \neq 0 \\ s \neq |M|}} \frac{(s-1)!(|M|-s-1)!}{(|M|-1)!}(v^M_{\text{imm}}(S) - x(S))^2. \quad (6)$$

Here $v^M_{\text{imm}}(S)$ is the restriction of the function v_{imm} on M. The minimum of the right-hand side of (6), according to Theorem 2, is attained on the Shapley vector of the finite game obtained by the restriction of the function v_{imm} on M. The monotonicity of v implies the monotonicity of v^M_{imm}. Therefore this minimum is attained at some imputation; since N is the carrier of the restriction of v_{imm} on M, this imputation is concentrated on N. Now due to the density of $X_f(v_{\text{imm}})$ in $X(v_{\text{imm}})$ we see that for each imputation x such that $v_{\text{imm}} - x \in L_2(2^J, \mu)$ and for each $\varepsilon > 0$ there is a finite imputation concentrated on N such that

$$\int_{2^J}(v_{\text{imm}}(S) - x^N(S))^2 d\mu(S) \leq \\ \leq \int_{2^J}(v_{\text{imm}}(S) - x(S))^2 d\mu(S) + \varepsilon.$$

Thus, the minimum of the functional (1) is attained on the set of imputations concentrated on N. Therefore, due to Theorem 2 and formula (6), for $M = N$ the minimum of the functional (1) is attained on the Shapley vector of a finite game given by the characteristic function v. This completes the proof. □

Theorem 5 *If the Shapley imputation exists then it is unique.*

Proof. Suppose that the minimum of the functional (1) is attained at two points x_1 and x_2. Then the measure of the set where

$$\left(v(S) - \frac{x_1(S) + x_2(S)}{2}\right)^2 <$$

$$< \frac{(v(S) - x_1(S))^2}{2} + \frac{(v(S) - x_2(S))^2}{2} \tag{7}$$

is positive, since this set is equal to the set of points in which $x_1(S) \neq x_2(S)$, and the later is open in the Tychonoff product topology. The strict inequality follows from the convexity of the functions $(a - t)^2$. Integrating inequality (7) over the measure μ we obtain a contradiction, which completes the proof. □

Acknowledgement This chapter is the author's translation of the original paper in: Teoretiko-igrovye voprosy prinyatia reshenij (Game-theoretic problems of decision-making), Leningrad, Nauka, 1978, 4–9.

References

1. Aumann, R.J., and L.S. Shapley (1976): *Values of non-atomic games.* Princeton, Princeton University Press.
2. Keane, M. (1969): *Some topics in N-person game theory.* Thesis, Northwestern University, Evanston, Illiois.

Chapter 7

Cores and generalized NM-solutions for some classes of cooperative games

V.A. Vasil'ev

Abstract: In this paper consisting of three parts, interconnection is analyzed between nonsymmetric analogs of the Shapley value, core allocations, and generalized von Neumann-Morgenstern solutions for both TU-cooperative games and NTU-games, associated with pure exchange economies. Strong attention is paid to the accessibility of the cores via a dynamic process of subsequent improvements of dominated alternatives, as well as to the characterization of undominated outcomes in terms of Harsanyi imputations and market equilibrium mechanisms.

In the first part we elaborate a unified functional approach to the investigation of the Harsanyi sets, cores and some nonsymmetric values of TU-cooperative games. The second part deals with the existence and classification problems for several types of generalized von Neumann-Morgenstern solutions, being realizations of a new optimality principle, combining the ideas of dynamics and stability. One of the main problems here relates to the construction of dynamic systems, whose final sets are globally stable and coincide with the cores of the games under consideration, while their transition correspondences generate merely dominating imputations of these games. In the third, concluding part of the paper, some results of a game-theoretic analysis of the totally contractual allocations in pure exchange models are given. Rather simple sufficient conditions, providing coincidence of the totally contractual core and the set of Walrasian equilibrium allocations are established, and the structure of domination relations, induced by several rules of entering and breaking contract systems, is studied.

Key words: Totally positive game, Harsanyi set, core, von Neumann-Morgenstern solution, totally contractual set, equilibrium allocation.

1 Introduction

This paper, consisting of three parts, is mostly devoted to the comparative analysis of some classes of nonsymmetric values, core allocations, generalized von Neumann-Morgenstern solutions and totally-contractual sets. In addition to the new results, the paper contains an extended survey of previously published articles by the author, related to the problems in question. Most of the main results, included in this survey, are provided with new proofs, based on rather general and efficient approaches that are of interest in their own right. It is worth to mention that these proofs are, certainly, not the shortest ones. However, they seem to be rather instructive and can easily be extended to more general settings.

In the first part of the paper a unified functional approach, based on the vector lattice theory, is applied to the investigation of the nonsymmetric values, analogous to the well-known Shapley value. A rather natural notion of the totally positive game, introduced by the author (see, e.g., Vasil'ev, 1975), made it possible to propose a new axiomatization of the Shapley value, treated as a positive symmetric linear operator on the appropriate KB-space (Vasil'ev, 1975). Recall that any KB-space enjoys remarkable characteristics, including norm and order completeness, nice projection and extension properties, etc. (for more details on this kind of partially ordered vector spaces, known also as Kantorovich-Banach vector lattices, see, e.g., Aliprantis and Border, 1994). The properties noted proved to be very useful in studying the Shapley value, especially in case of the infinite cooperative games (see Vasil'ev, 1975; Vasil'ev, 1982; and Vasil'ev, 1998). Moreover, by applying the new type of axiomatization mentioned above, the concepts of so-called Harsanyi operators and Harsanyi set were elaborated in Vasil'ev (1975), Vasil'ev (1980), and Vasil'ev (1982) (see, also, Vasil'ev, 1981a). The Harsanyi set, consisting (by definition) of all possible distributions of the Harsanyi dividends (Harsanyi, 1959) of the coalitions among their members, was developed initially in order to provide a unified description of the core imputations in terms of some decentralization mechanisms. Surprisingly, this set turned out to be of a core-type structure itself. Namely, it is shown that the Harsanyi set coincides with the core of a properly defined convex game (Theorem 2.11). This is one of the main results of the paper, first established by Vasil'ev (1980) (see also Vasil'ev, 1981a). To single out some important consequences of this result, we first show the external stability of the Harsanyi imputation set under the most general assumptions (Theorem 2.18). One more noteworthy corollary of Theorem 2.11, exhibiting a rather interesting property of the extreme points of the Harsanyi set, shows that any such an extreme point is generated according to some ranking (permutation) π of the players, by assigning a positive Harsanyi dividend of a coalition to the player in the coalition being the π-last member, and a negative dividend of a coalition to the player being the π-first member (Vasil'ev, 1980).

The unified functional approach proposed in the first part proved to be helpful not only in the description of the Harsanyi set itself, but also in characterizing the cores of the almost positive and convex games in terms of the Harsanyi payoff vectors as well (Theorems 2.13 and 2.16). Note, that besides the vector lattice theory, this approach strongly relies upon the original integral representation of the support function of the core of a convex game (Theorem 2.4), which may be of interest in many other respects.

The second part of the paper is mainly addressed to issues related to the so-called generalized von Neumann-Morgenstern solutions (shortly gNM-solutions). These new solutions, introduced by the author, are based on the principle of sequential improvements of dominated alternatives and differ from the classic NM-solutions only by the fact that some relaxed type of external stability is applied. This relaxed type of external stability is chosen to be an external stability w.r.t. rather natural extensions of the initial domination relation α (to clarify an interpretation of the concepts given below, recall that $(x, y) \in \alpha$ means "y is as at least as good as x"). One of the main extensions is the well-known transitive closure α_* (i.e., (x, y) belongs to the binary relation α_* if and only if there exists a finite sequence $x_0, x_1, \ldots, x_m$ with $x_0 = x$, $x_m = y$ and $(x_r, x_{r+1}) \in \alpha$ for any $r = 0, 1, \ldots, m-1$). Roughly speaking, y improves (dominates) x w.r.t. α_* iff y results as an outcome of a finite number of sequential "α-improvements" starting at x. The notion of NM-rank, characterizing the number of improvements required to arrive at gNM-solution starting at the "most distant point", is introduced, and some gNM-existence theorems with evaluation of NM-rank are established. It is shown, that in contrast to the classic NM-solution, generalized NM-solutions for TU-cooperative games always exist. Moreover, for any cooperative game, transferable or nontransferable, the common part of these solutions equals the core, and in case the latter is empty the NM-rank of the game, defined as the minimal NM-rank of its gNM-solutions, is less or equal to three (Theorem 3.11). It is worth to note in this respect that for the famous Lucas 10-person game, which has no NM-solutions, it holds that the core is one of its gNM-solutions with NM-rank equals 2. Therefore, NM-rank of Lucas game itself is also equal to two (for more details see Vasil'ev, 1981b).

Besides the standard extensions, like transitive closure, we introduce a new concept of so-called limit transitive closure α^* by allowing the infinite convergent α-monotone sequences (trajectories): $(x, y) \in \alpha^*$ iff there exists an α-monotone convergent sequence $\{x_r\}_{r=0}^{\infty}$ with $x_0 = x$, $\lim x_r = y$, where as usual the sequence $\{x_r\}_{r=0}^{\infty}$ is called α-monotone if $(x_r, x_{r+1}) \in \alpha$ for any $r = 0, 1, \ldots$. A general topological approach, elaborated in the paper to deal with the limit transitive closure, is basically inspired by the well-known Mashler-Peleg theory of dynamic systems (Maschler and Peleg, 1976). Within the framework of this topological approach, it is proven that for any balanced TU-cooperative game there exists a dynamic system, whose final set is globally stable and equals the core, while outside the core all the outcomes of transition function dominate (improve) a current imputation. As a consequence the most important theorem of this part of the paper is established: for any imputation x outside the core $C(v)$ of a balanced TU-cooperative game v there exists a monotone trajectory of imputations starting at x and converging to some point belonging to the core $C(v)$ (Theorem 3.22). Combining this theorem with that obtained for the games with empty core, we get one of the main results of the paper: for any TU-cooperative game there exists an NM-solution w.r.t. the limit transitive closure of the classic domination relation. This solution is unique and coincides with the core, if the latter is not empty, and can be chosen to contain a finite number of imputations otherwise (Theorem 3.24). To show that the TU-property is essential, we construct an example of an NTU-cooperative game G with a "black hole", being a closed set $B \subseteq G(N)$ that does not intersect the core $C(G)$ and contains all the monotone trajectories originating at any point $x \in B$.

The third, concluding part of the paper, contains a game-theoretical analysis of the totally contractual allocations, similar to that introduced by Makarov (1982) in order to describe stable outcomes of some quite natural recontracting processes in pure exchange economies. The structure of domination relations, induced by several rules of entering and breaking contract systems, is studied, and rather simple conditions, providing the coincidence of the totally contractual core and Walrasian equilibria are established (Theorem 4.27). One of the main distinctive features of the recontracting processes under consideration is the almost inevitable multiplicity of outcomes of breaking procedures. This lack of uniqueness is caused by the very weak restrictions, imposed on the systems of deliberately nullified contracts. In particular, it may often be the case that after by some coalition chosen contracts are broken, the rest (including newly concluded by this coalition) do not constitute a feasible system. It is inherent in the model that in such a situation the breaking process proceeds spontaneously (just as it happens in the real life), and stops after feasibility of the contract system is recovered. The only requirement we deal with on this step is minimality: the spontaneous process cancels (nullifies) collections of contracts as small as possible, provided that not nullified ones constitute a feasible contract system. Since we are looking for systems that are stable with respect to any possible outcome of the breaking procedure, the above mentioned multiplicity makes it rather difficult to characterize these stable contract systems. Therefore, to obtain new core equivalence results (like Theorem 4.27 and its corollaries), characterizing equilibrium allocations in terms of quasi-stable contract systems, we apply an appropriate reduction of the contractual blocking to a more simple domination relation in NTU-cooperative games, associated with the recontracting procedure in question (Propositions 4.8, 4.9, and 4.11).

2 H-Payoffs and the cores of TU-cooperative games

Recall first the definition of an n-person TU-cooperative game, which is prescribed entirely by its characteristic function form, given by a real-valued characteristic function $v : 2^N \to \Re$, where $N := \{1, \cdots, n\}$, and 2^N is the collection of all subsets of N. Following the terminology adopted (see, e.g., Aumann and Shapley, 1974; von Neumann and Morgenstern, 1953; and Rosenmuller, 1981), each element of the set N is called a player. Further, each subset S of the set N is called a coalition, and a value $v(S)$ is interpreted as the maximal (aggregated) income, available to the coalition S. Denote by $I(v)$ the set of imputations (individually rational efficient payoffs) of the TU-cooperative game v (to simplify some formulas, we denote a singleton $\{i\} \subseteq N$ by i when no confusion can arise):

$$I(v) := \{x \in \Re^N \mid x(N) = v(N),\ x_i \geq v(i),\ i \in N\},$$

where, applying the standard shortening, we denote $x(S) := \sum_{i \in S} x_i$ for any $x \in \Re^N$ and $S \subseteq N$. In the sequel we also denote $\mathcal{C} := 2^N \setminus \{\emptyset\}$.

We say that an imputation $x \in I(v)$ is dominated by an imputation $y \in I(v)$, if there exists a coalition $S \in \mathcal{C}$ such that (i) $x_i < y_i$, for any $i \in S$, and (ii) $y(S) \leq v(S)$. We denote this domination relation by α_v and define

$$C(\alpha_v) := \{x \in I(v) \mid x \text{ is not dominated by any } y \in I(v)\}.$$

Recall (see e.g. Rosenmuller, 1981), that the set $C(\alpha_v)$ of the maximal (w.r.t. the binary relation α_v) elements of the set $I(v)$ is called the core of cooperative game v.

The aim of this part of the paper is to present a characterization of the cores of TU-cooperative games in terms of nonsymmetric analogs of the Shapley value. It is worth to note, that the principle of optimality, corresponding to the Shapley value, has nothing in common (at least, in the finite case) with the coalitional stability concept, on which the notion of the core is based. Nevertheless, as it will be shown in the next subsections, the characterization mentioned above turns out to be possible for the rather large class of TU-cooperative games.

2.1 Harsanyi Payoff Vectors

We identify the set consisting of all n-person TU-cooperative games with the function space $V = V(N)$ of the mappings $v : 2^N \to \Re$, satisfying the requirement $v(\emptyset) = 0$. Let v be an arbitrary TU-cooperative game (shortly c.g.) belonging to V. For any $v \in V$, we define the values v_T (so-called Harsanyi dividends (Harsanyi, 1959) of the c.g. v) as the components of the (unique) solution of the system of linear equations

$$v(S) = \sum_{T \subseteq S} v_T, \quad S \subseteq N. \tag{2.1}$$

Recall (see Rosenmuller, 1981) that the Shapley value of the c.g. v is the vector $\Phi^o(v) \in \Re^N$, defined by

$$\Phi^o(v)_i := \sum_{T \in \mathcal{C}_i} \frac{1}{|T|} v_T, \quad i \in N, \tag{2.2}$$

where, as usual, $|T|$ denotes the cardinality of the (finite) set T, and

$$\mathcal{C}_i := \{T \in \mathcal{C} \mid i \in T\}.$$

We interpret the values v_T as (aggregated) gains, obtained by the "unions" T as a result of their formation (aggregated expenses, spent to support the "unions" T in case $v_T < 0$). Formula (2.1) expresses the fact that the maximal (aggregated) income available to the coalition S, equals to the total gain, delivered to the coalition S by all the "unions" potentially realized within this coalition. Hence the Shapley value describes the imputation resulting from dividing equally the dividends v_T of the "unions" T amongst their members; the total gain earned by the player $i \in N$ is equal to the sum of all incomes obtained from all the "unions" listing her as their member.

Also recall (see e.g. Rosenmuller, 1981, or Vasil'ev, 1984) that the Harsanyi dividends of a c.g. $v \in V$ can be found as the coefficients in the expression $v = \sum_{T \in \mathcal{C}} v_T \epsilon^T$ of v w.r.t. the well-known basis of the vector space V consisting of the so-called unanimity games ϵ^T, $T \in \mathcal{C}$, defined by the formulas

$$\epsilon^T(S) := \begin{cases} 1, & \text{for all } S, \text{ such that } T \subseteq S, \\ 0, & \text{otherwise.} \end{cases}$$

Note, that for any $T \in \mathcal{C}$ it holds that the dividend $(\epsilon^T)_S$ of coalition S in unanimity game ϵ^T is given by

$$(\epsilon^T)_S := \begin{cases} 1, & \text{if } S = T, \\ 0, & \text{otherwise.} \end{cases}$$

Following Vasil'ev (1981a), we introduce the cone of so-called *totally positive* TU-cooperative games (the first, more general definition of a totally positive TU-cooperative game, suitable for both finite and infinite number of players, was given in Vasil'ev, 1975) by

$$V_+ := \{v \in V \mid v_T \geq 0,\ T \subseteq N\}.$$

Further, denote by $\mathcal{H}$ the class of linear operators $\Phi : V \to \Re^N$, satisfying the conditions

$$\Phi(V_+) \subseteq \Re^N_+, \tag{2.3}$$
$$\Phi(v)(T) = v(N),\ v \in V,\ T \in \text{Supp } v, \tag{2.4}$$

where $\text{Supp}\, v := \{T \subseteq N \mid v(S \cap T) = v(S),\ S \subseteq N\}$. It is easy to see, that the linear operator Φ^o, defined by the formula (2.2), belongs to the family $\mathcal{H}$, and, besides, satisfies the symmetry condition

$$\Phi^o(\pi \circ v) = \pi \circ \Phi^o(v),\ v \in V,\ \pi \in \Pi, \tag{2.5}$$

where $\Pi = \Pi(N)$ is the set of all permutations (one-to-one maps) of N, and $\pi \circ v(S) := v(\pi(S))$, $\pi \in \Pi$. Moreover, the linear operator Φ^o is uniquely determined by the properties (2.3)–(2.5) (the same is true under the appropriate refinement for the regular infinite games as well, Vasil'ev, 1975).

By virtue of the properties Φ^o, mentioned above, the operators belonging to the collection $\mathcal{H} \setminus \{\Phi^o\}$ are treated as nonsymmetric analogs of the operator Φ^o, and that is why the Harsanyi payoff vectors, based on these operators (see the formal definition below), are considered as nonsymmetric analogs of the Shapley value.

Definition 2.1 Elements of the set $H(v) := \{\Phi(v) \mid \Phi \in \mathcal{H}\}$ are called Harsanyi payoff vectors (shortly H-payoffs) of a c.g. v.

Define

$$\mathcal{P} := \{[p_i^T]_{i\in N}^{T\in\mathcal{C}_i} \mid \sum_{i\in T} p_i^T = 1,\ p_i^T \geq 0, \quad T \in \mathcal{C},\ i \in T\}$$

and, for any $p = [p_i^T]_{i\in N}^{T\in\mathcal{C}_i} \in \mathcal{P}$, define the operator $\Phi_p : V \to \Re^N$ by the formula

$$\Phi_p(v)_i := \sum_{T\in\mathcal{C}_i} p_i^T v_T, \quad i \in N,\ v \in V.$$

Then it can easily be shown (see Vasil'ev, 1982; Vasil'ev, 1984), that the following representation theorem holds.

Theorem 2.2 *Operator Φ belongs to $\mathcal{H}$ iff there exists $p \in \mathcal{P}$ such that $\Phi = \Phi_p$.*

As a consequence of Theorem 2.2 it follows that to obtain an H-payoff of the c.g. v one can apply an element $p \in \mathcal{P}$ as a sharing mechanism to distribute the dividends amongst the players, similar as in forming the Shapley value. The only difference is that in a general situation the participants of the unions T get shares of the corresponding dividends v_T which are not necessarily equal to $\bar{p}_i^T := \frac{1}{|T|}$. An elements $p = [p_i^T]_{i\in N}^{T\in\mathcal{C}_i} \in \mathcal{P}$ defines a concrete realizations of the sharing mechanism, and the set $H(v)$ describes the collection of all the outcomes of these realizations.

2.2 Dual description of the set of H-payoffs

Before we consider the dual description of the polyhedrons $H(v)$ in terms of linear inequalities, let us recall that the so-called *c*-core of a c.g. v is defined by

$$C(v) := \{x \in I(v) \mid x(S) \geq v(S),\ S \subseteq N\}.$$

It is clear that $C(v) \subseteq C(\alpha_v)$ for any $v \in V$. Observe also that it can easily be shown that the coincidence of the core and c-core takes place iff v meets the requirement

$$v(N) \geq v(S) + \sum_{N \setminus S} v(i), \quad S \subseteq N. \tag{2.6}$$

In this subsection it will be established that the set of the H-payoffs of any c.g. v can be represented as the c-core of some associated c.g. v_H to be defined below.

To describe these associated games v_H, we introduce some relevant concepts and notations. First, by means of the convex cone V_+ of totally positive games we define the partial order $\leq_o$ in the vector space V:

$$v \leq_o u \Leftrightarrow u - v \in V_+.$$

Following the vector lattice terminology, we call functions (games) $v^+ := v \vee 0$, $v^- := (-v) \vee 0$, and $|v| := (-v) \vee v$ a positive, negative, and total variation of v, respectively ($\vee$ denotes the supremum sign in the vector lattice $(V, \leq_o)$).

Remark 2.3 One can easily verify, that the Harsanyi dividends of v^+, v^-, and $|v|$ are defined (in terms of the dividends of v) as follows: $v^+_T = (v^+)_T := \max\{v_T, 0\}$, $v^-_T = (v^-)_T := \max\{-v_T, 0\}$, and $|v|_T = (|v|)_T := \max\{-v_T, v_T\}$, for any $T \in \mathcal{C}$.

In the notations introduced above the functions v_H can be defined as follows:

$$v_H(S) := v(S) - v_2^-(S, N \setminus S), \quad S \subseteq N,$$

where $v_2^-(S, N \setminus S) := v^-(N) - v^-(S) - v^-(N \setminus S)$.

The key role in the proof of the equality

$$H(v) = C(v_H) \tag{2.7}$$

is played by the representation theorem providing a description of the support function of the core of a convex game v in terms of the corresponding Harsanyi dividends v_T. Recall from Rosenmuller (1981) that c.g. v is convex, if for any $S, T \subseteq N$ the inequality

$$v(S \cup T) + v(S \cap T) \geq v(S) + v(T)$$

holds. We denote the support function of the core $C(\alpha_v)$ by h_v. Recall from Rockafellar (1970) that, directly by definition of the support function of a convex set, we have the following formula

$$h_v(q) := \sup\{q \cdot x \mid x \in C(\alpha_v)\}, \quad q \in \Re^N$$

(here and below $q \cdot x$ is the inner product of the vectors q and x).

Theorem 2.4 *For any convex game v the support function h_v can be represented as follows*

$$h_v(q) = \sum_{T \in \mathcal{C}} v_T \max\{q_i \mid i \in T\}, \quad q \in \Re^N. \tag{2.8}$$

Proof. Let q be an arbitrary element of $\Re^N$. Denote by $r(q)$ the number of distinct values, admitted by the components of q (i.e., $r(q) := |\{q_k \mid k \in N\}|$), and set $T_0 := \emptyset$, $T_k := \{i \in N \mid q_i \leq \min_{j \in N \setminus T_{k-1}} q_j\}$, $k = 1, \ldots, r(q)$.

Let $v \in V$ be any convex c.g. It is clear, that convex games satisfy requirement (2.6). Therefore the core of the game v coincides with its c-core and, hence, $C(\alpha_v) = C(v)$ is a bounded convex polyhedron. Consequently, by definition of the support function h_v, there exists an imputation $x_q \in C(v)$ satisfying the equality $q \cdot x_q = h_v(q)$. Set $\mathcal{T}_q := \{S \mid x_q(S) = v(S)\}$ and prove that $T_k \in \mathcal{T}_q$ for any $k = 1, \ldots, r(q)$. Observe first, that $\emptyset$, $N \in \mathcal{T}_q$ and, besides, $\mathcal{T}_q$ is closed w.r.t. intersection and union of any two elements of $\mathcal{T}_q$. Indeed, let S_1, S_2 belong to $\mathcal{T}_q$. Due to the convexity of v we have $v(S_1 \cup S_2) + v(S_1 \cap S_2) \geq x_q(S_1) + x_q(S_2)$. Hence, taking into account the inclusion $x_q \in C(v)$ and the identity $x_q(S_1 \cup S_2) + x_q(S_1 \cap S_2) = x_q(S_1) + x_q(S_2)$, we get the required $S_1 \cup S_2$, $S_1 \cap S_2 \in \mathcal{T}_q$. We now prove that $T_k \in \mathcal{T}_q$ for any $k = 1, \ldots, r(q)$. To do so, fix some $k \leq r(q)$ and consider a collection $\mathcal{T}_q^k := \{S \in \mathcal{T}_q \mid T_k \subseteq S\}$. Because of the closedness of $\mathcal{T}_q$ w.r.t. finite intersections, the collection $\mathcal{T}_q^k$ contains the minimal (w.r.t. inclusion) element, say S_k. We show that $S_k = T_k$. Since $T_{r(q)} = N$, we may assume w.l.g. that $k < r(q)$. Suppose, $S_k \neq T_k$. Choose some $i \in S_k \setminus T_k$ and for every $j \in T_k$ define

$$\mathcal{T}_{ij} := \{S \subseteq N \mid i \notin S,\, j \in S\},\ \delta_{ij} := \min\{x_q(S) - v(S) \mid S \in \mathcal{T}_{ij}\}.$$

Two cases may occur: 1) $\forall j \in T_k$ it holds that $\delta_{ij} = 0$ and 2) $\exists j_o \in T_k$ such that $\delta_{ij_o} > 0$. If the first case holds, choose arbitrary coalitions $S_{ij} \in \mathcal{T}_{ij} \bigcap \mathcal{T}_q$, $j \in T_k$, and put $T_k^i := S_k \bigcap \left(\bigcup_{j \in T_k} S_{ij}\right)$. Due to the closedness of the collection $\mathcal{T}_q$ w.r.t. finite intersections and unions of its members, we have that $T_k^i \in \mathcal{T}_q$. But by construction it holds: $T_k \subseteq T_k^i \subseteq S_k$ and $i \notin T_k^i$, which contradicts the minimality of S_k. If the second case prevails, choose an arbitrary $\varepsilon \in (0, \delta_{ij_o})$ and put $x_q^\varepsilon := x_q + \varepsilon e^i - \varepsilon e^{j_o}$, where e^i, e^{j_o} are the corresponding (unit) basis vectors of $\Re^N$. It follows from the definition of x_q^ε that

$$x_q^\varepsilon(S) = \begin{cases} x_q(S), & i, j_o \in S, \\ x_q(S) + \varepsilon, & i \in S,\, j_o \notin S, \\ x_q(S) - \varepsilon, & i \notin S,\, j_o \in S. \end{cases}$$

Hence, due to the choice of ε, we get: $x_q^\varepsilon \in C(v)$. But since $q_i > q_{j_o}$, the inequality $q \cdot x_q^\varepsilon > q \cdot x_q$ holds. As the latter contradicts to the assumed equality $q \cdot x_q = \sup\{q \cdot x \mid x \in C(v)\}$, we get $S_k = T_k$. Due to the arbitrariness of k, the latter equality means that the required inclusions $T_k \in \mathcal{T}_q$ hold for any $k = 1, \ldots, r(q)$.

We now turn to the proof of the formula (2.8) itself. From the equalities $x_q(T_k) = v(T_k)$ and the definition of T_k we obtain that $q \cdot x_q = \sum_{k=1}^{r(q)} \bar{q}_k(v(T_k) - v(T_{k-1}))$, where $\bar{q}_k := \max\{q_i \mid i \in T_k\}$. Further, by definition of v_T we have $v(T_k) - v(T_{k-1}) = \sum_{T \in T_{k-1,k}} v_T$ with $T_{k-1,k} := \{T \subseteq T_k \mid T \cap (T_k \setminus T_{k-1}) \neq \emptyset\}$. Taking into account that $\max\{q_i \mid i \in T\} = \bar{q}_k$ for any $T \in T_{k-1,k}$, by the relations given above we obtain the required formula

$$h_v(q) = q \cdot x_q = \sum_{T \in \mathcal{C}} v_T \max\{q_i \mid i \in T\}. \qquad \square$$

As a straightforward corollary of the inclusions $T_k \in \mathcal{T}_q$, $k = 1, \ldots, r(q)$, established above, we may give a rather simple proof of the famous Shapley Theorem on the extreme points of the core of a convex TU-game (Shapley, 1971). Below, for any permutation $\pi = (i_1, \ldots, i_n) \in \Pi$ and function $v \in V$ we denote by v^π the vector $v^\pi := (v_1^\pi, \ldots, v_n^\pi)$, where $v_k^\pi := v(S_k^\pi) - v(S_{k-1}^\pi)$, $S_0^\pi := \emptyset$, and $S_k^\pi := \{i_1, \ldots, i_r\}$ with $i_r = k$, $k = 1, \ldots, n$.

Theorem 2.5 *The set* $\operatorname{ex} C(v)$ *of extreme points of the core of a convex game* v *has the following representation*

$$\operatorname{ex} C(v) = \{v^\pi \mid \pi \in \Pi\}.$$

Proof. Since $C(\alpha_v) = C(v)$ is a bounded convex polyhedron in finite-dimensional vector space, we have: vector x belongs to $\operatorname{ex} C(v)$ iff there exists some open set $G^x \subseteq \Re^N$ such that for any element $z \in G^x$ the subset $C_v^z := \{y \in C(v) \mid h_v(z) = z \cdot y\}$ of $C(v)$ contains x as its unique element, thus $C_v^z = \{x\}$. But any open set in $\Re^N$ contains a nonempty subset of vectors with pairwise distinct components. Consequently, if x belongs to $\operatorname{ex} C(v)$, then there exists $q \in \Re^N$ such that $C_v^q = \{x\}$, and $q_{i_1} < q_{i_2} < \ldots < q_{i_n}$ for some $\pi = (i_1, i_2, \ldots, i_n) \in \Pi$. Moreover, due to the equalities $T_k = S_{i_k}^\pi$ it holds that $x(S_k^\pi) = v(S_k^\pi)$ for any $k = 1, \ldots, n$. But then, obviously, the equality $x = v^\pi$ holds.

Reversely the inclusion $\{v^\pi \mid \pi \in \Pi\} \subseteq \operatorname{ex} C(v)$ can be checked directly.
$\square$

To mention one more straightforward consequence of Theorem 2.4, following immediately from the Duality Theorem for linear programming, for any

$q \in \Re^N$ set $\Lambda_q := \{(\lambda_T)_{T\in\mathcal{C}} \in \Re^{\mathcal{C}} \mid \lambda_N e^N - \sum_{T\in\mathcal{C}\setminus\{N\}} \lambda_T e^T = q,\ \lambda_T \geq 0,\ T \in \mathcal{C}\setminus\{N\}\}$, where components of the vectors $e^T \in \Re^N$ are defined as follows: $(e^T)_i = 1$, if $i \in T$, and $(e^T)_i = 0$, if $i \in N \setminus T$.

Corollary 2.6 *For any convex game $v \in V$ it holds that*

$$\min\{\lambda_N v(N) - \sum_{T\in\mathcal{C}\setminus\{N\}} \lambda_T v(T) \mid (\lambda_T)_{T\in\mathcal{C}} \in \Lambda_q\}$$
$$= \sum_{T\in\mathcal{C}} v_{_T} \max\{q_i \mid i \in T\}, \quad q \in \Re^N.$$

To conclude the discussion on Theorem 2.4, note that its analogs may be of considerable interest in case of an infinite number of players as well. Some generalizations of the formula (2.8) to the case of infinite convex game on a metric compactum can be found in Vasil'ev and Zuev (1988).

Next we prove that v_H is a convex game for any $v \in V$. Therefore, for any collection of coalitions S_i, $i = 1,\ldots,m$, and for any $v \in V$, we define

$$[S_1,\ldots,S_m] := \Big\{T \subseteq \bigcup_{i=1}^m S_i \mid T \cap S_i \neq \emptyset, \quad i = 1,\ldots,m\Big\},$$
$$v_m(S_1,\ldots,S_m) := \sum_{\omega\subseteq\{1,\ldots,m\}} (-1)^{m-|\omega|} v(\bigcup_{i\in\omega} S_i).$$

From the definition of convex game we get immediately the following convexity test:
Function $u \in V$ is convex iff for any pairwise disjoint coalitions S_1, S_2, S_3 the inequality

$$u_2(S_1,S_2) + u_3(S_1,S_2,S_3) \geq 0 \tag{2.9}$$

holds.
Applying induction on m, one can easily prove that for any collection of pairwise disjoint nonempty coalitions $S_1,\ldots,S_m$ it holds that

$$\sum_{T\in[S_1,\ldots,S_m]} (-1)^{|T|} = (-1)^m, \tag{2.10}$$

and, hence, for such a collection of coalitions and for any $v \in V$ the equality

$$v_m(S_1,\ldots,S_m) = \sum_{T\in[S_1,\ldots,S_m]} v_{_T} \tag{2.11}$$

holds. To make use of formula (2.11) and criterion (2.9) for proving the convexity of v_H, we express the Harsanyi dividends of this game in terms of the positive and negative variations of the game v. Therefore we first calculate the dividends of the game $v^{(2)}(S) := v_2(S, N \setminus S)$, $S \subseteq N$.

Lemma 2.7 *For any game $v \in V$ the Harsanyi dividends $(v^{(2)})_T$ of the game $v^{(2)}$ are as follows*

$$(v^{(2)})_T = -v_T + (-1)^{|T|+1} \sum_{R \subseteq N \setminus T} v_{T \cup R}, \;\; T \in \mathcal{C}.$$

Proof. Follows by induction on $k = |T|$, based on the relations (2.1) and (2.10). □

Corollary 2.8 *For any game $v \in V$ and for any coalition $T \in \mathcal{C}$ it holds*

$$(v_H)_T = v_T^+ + (-1)^{|T|} \sum_{R \subseteq N \setminus T} v_{T \cup R}^-. \tag{2.12}$$

Proof. Follows from the representation $v_H = v - (v^-)^{(2)}$ and Lemma 2.7.□

Proposition 2.9 *The function v_H is convex for any c.g. $v \in V$.*

Proof. In order to apply criterion (2.9) to the function v_H, consider an arbitrary collection of pairwise disjoint coalitions S_1, S_2, S_3, and fix some element v_R^- with $R_i \neq \emptyset$, $i = 1, 2$, where $R_i := R \bigcap S_i$, $i = 1, 2, 3$. Making use of Corollary 2.8, let us calculate the number c_R of entrances of the term v_R^- in the left-hand side of the inequality (2.9) with $u = v_H$. It is rather easy to verify that

$$c_R = \begin{cases} \sum_{T \in [R_1, R_2]} (-1)^{|T|} + \sum_{T \in [R_1, R_2, R_3]} (-1)^{|T|} &, \quad R_3 \neq \emptyset, \\ \sum_{T \in [R_1, R_2]} (-1)^{|T|} &, \quad R_3 = \emptyset. \end{cases}$$

From here, due to (2.10) we get $c_R = 0$ if $R_3 \neq \emptyset$, and $c_R = 1$ if $R_3 = \emptyset$. Since by definition of the negative variation v^- the values v_R^- are nonnegative, we obtain indeed that $u = v_H$ meets the requirement (2.9). □

Let us give one more auxiliary result, which is relevant to prove equality (2.7). This result represents some generalization of formula (2.10) and may be interesting in itself.

Lemma 2.10 *For any collection of pairwise disjoint nonempty subsets S_1, ..., $S_m \subseteq N$, and for any $q \in \Re^N$ it holds*

$$\begin{aligned} &\sum_{T \in [S_1, \ldots, S_m]} (-1)^{|T|} \max\{q_i \mid i \in T\} \\ = \;\; &(-1)^m \max_{j \in \{1, \ldots, m\}} \min\{q_i \mid i \in S_j\}. \end{aligned} \tag{2.13}$$

Proof. By induction on m. Let $m = 1$. W.l.g. we may assume that $S_1 = N$. Take c.g. $u \in V$ with $u_T = (-1)^{|T|}$, $T \in \mathcal{C}$. Applying convexity criterion (2.9), it is easy to see that c.g. u thus defined is convex. In fact, an elementary calculation shows, that for the game u in question the left-hand side of the inequality (2.9) takes the form

$$\sum_{T\in[T_1,T_2]} (-1)^{|T|} + \sum_{T\in[T_1,T_2,T_3]} (-1)^{|T|}.$$

But due to the equality (2.10) the latter expression is always nonnegative. Hence, the game u is convex, and by Theorem 2.4 we get

$$h_u(q) = \sum_{T\in\mathcal{C}} (-1)^{|T|} \max\{q_i \mid i \in T\}. \tag{2.14}$$

On the other hand, because of the equalities $u(S) = \sum_{T\subseteq S} u_T$ and (2.10) we obtain $u(S) = -1$ for any $S \neq \emptyset$. Consequently, as it can easily be shown, we get

$$C(\alpha_u) = \{x \in \Re^N \mid x(N) = -1,\ x_i \leq 0,\ i \in N\}.$$

But $\sup\{q \cdot x \mid x \in C(\alpha_u)\} = -\min\{q_i \mid i \in N\}$, which completes (due to the equality (2.14)) the proof of our lemma for the case $m = 1$.

Suppose the equality (2.13) is valid for all $m \leq r$ and consider an arbitrary collection of nonempty pairwise disjoint coalitions $S_1, \ldots, S_r, S_{r+1}$. W.l.o.g. we may assume that the inequalities

$$\min_{S_1} q_i \leq \ldots \leq \min_{S_r} q_i \leq \min_{S_{r+1}} q_i$$

hold. Put $q[S_1, \ldots, S_k] := \sum_{T\in[S_1,\ldots,S_k]} (-1)^{|T|} \max\{q_i \mid i \in T\}$. From the obvious equality

$$\begin{aligned} & q[S_1, \ldots, S_r, S_{r+1}] \\ = \ & q[S_1, \ldots, S_{r-1}, S_r \cup S_{r+1}] - \sum_{k=r}^{r+1} q[S_1, \ldots, S_{r-1}, S_k], \end{aligned}$$

we get (on the base of inductive hypothesis):

$$q[S_1, \ldots, S_r, S_{r+1}] = (-1)^{r+1} \min\{q_i \mid i \in S_{r+1}\}.$$

It is clear that this equality completes the proof of Lemma 2.10. □

Theorem 2.11 *The set of Harsanyi payoffs of any c.g. $v \in V$ coincides with the c-core of c.g. v_H.*

Proof. Since $H(v)$ and $C(v_H)$ are convex compact sets for any $v \in V$, to prove the theorem it is sufficient to establish the coincidence of support functions $h_u(q) = \sup\{q \cdot x \mid x \in C(u)\}$, with $u := v_H$ and $g_v(q) := \sup\{q \cdot x \mid x \in H(v)\}$. Taking into account the equality

$$H(v) = \sum_{T \in \mathcal{C}} v_T^+ A_T - \sum_{T \in \mathcal{C}} v_T^- A_T,$$

where $A_T := \{x \in \Re_+^N \mid x(T) = 1,\ x(N \setminus T) = 0\}$, it is not very hard to prove the formula

$$g_v(q) = \sum_{T \in \mathcal{C}} v_T^+ \max\{q_i \mid i \in T\} - \sum_{T \in \mathcal{C}} v_T^- \min\{q_i \mid i \in T\}.$$

As to the support function of the core of convex game $u = v_H$, by applying Corollary 2.8, Proposition 2.9, and Theorem 2.4 we obtain

$$\begin{aligned} h_u(q) &= \sum_{T \in \mathcal{C}} v_T^+ \max\{q_i \mid i \in T\} \\ &\quad + \sum_{T \in \mathcal{C}} \sum_{R \subseteq N \setminus T} (-1)^{|T|} v_{T \cup R}^- \max\{q_i \mid i \in T\}. \end{aligned}$$

Adding up the coefficients at some v_T^- in the right-hand side of the expression written above, we get the formula for their sum d_T:

$$d_T = \sum_{\emptyset \neq R \subseteq T} (-1)^{|R|} \max\{q_i \mid i \in R\}.$$

Due to Lemma 2.10 for any $T \in \mathcal{C}$ and $q \in \Re^N$ it holds that

$$\sum_{\emptyset \neq R \subseteq T} (-1)^{|R|} \max\{q_i \mid i \in R\} = -\min\{q_i \mid i \in T\}.$$

Thus, applying the latter equality, we obtain

$$\sum_{T \in \mathcal{C}} \sum_{R \subseteq N \setminus T} (-1)^{|T|} v_{T \cup R}^- \max\{q_i \mid i \in T\} = -\sum_{T \in \mathcal{C}} v_T^- \min\{q_i \mid i \in T\},$$

which completes the proof. □

Remark 2.12 Since v_H is a convex game for any $v \in V$, one of the main consequences of the equality $H(v) = C(v_H)$ exhibits a rather interesting property of the extreme points of the Harsanyi set $H(v)$. Namely, on the basis of the already mentioned Shapley Theorem (Theorem 2.5 above) it was shown by Vasil'ev (1980) that each extreme point of the Harsanyi set $H(v)$ is associated with some permutation $\pi \in \Pi$, by assigning to every player $i \in N$ any positive Harsanyi dividend v_S provided i is the π-last member of S, and any negative dividend v_T when this player is the π-first member of T (for more details, besides Vasil'ev, 1980, see also Vasil'ev, 2003). Recall (Theorem 2.5), that in contrast to the Harsanyi set, each extreme point v^π of the c-core $C(v)$ of a convex game v is generated by assigning to any player $i \in N$ each Harsanyi dividend v_S in case i is the π-last member of S (independently on weather dividend v_S is positive or negative).

2.3 H-Payoffs and the core

To start with a detailed characterization of the cores of TU-games, we first prove a direct corollary of Theorem 2.11. Denote by $v_{(1)}$ the additive component of c.g. v

$$v_{(1)}(S) := \sum_{i \in S} v(i), \quad S \subseteq N.$$

Theorem 2.13 *If c.g. v satisfies requirement (2.6), then any undominated imputation (i.e., core-imputation) of this game is an H-payoff vector. Moreover, independently on whether v meets the condition (2.6) or not, the equality $C(\alpha_v) = H(v)$ takes place iff $v - v_{(1)} \in V_+$.*

Proof. By definition of v_H we have $v_H(S) \leq v(S)$ for any $S \subseteq N$, and, besides, $v_H(N) = v(N)$. Consequently, Theorem 2.11 implies

$$C(v) \subseteq H(v) \tag{2.15}$$

for any $v \in V$. Because of (2.15), the first assertion of the theorem follows immediately from the equality $C(\alpha_v) = C(v)$, which is valid for any c.g. v satisfying condition (2.6).

As for the second assertion, letting $u = v - v_{(1)}$ and taking into account that $C(\alpha_v) = v_{(1)} + C(\alpha_u)$, $H(v) = v_{(1)} + H(u)$, it is enough to verify that for any $u \in V$ with $u_{(1)} = 0$ it holds: $C(\alpha_u) = H(u) \Leftrightarrow u \in V_+$. It is clear that all the core-imputations of any such a c.g. u are nonnegative vectors. Hence, assuming $H(u) = C(\alpha_u)$, by Theorem 2.2 we obtain $u \in V_+$. Indeed, if we assume the contrary, i.e., $u_T < 0$ for some $T \subseteq N$, we can choose such a $p \in \mathcal{P}$ that $p_i^T = 1$ for some $i \in T$, and $p_i^S = 0$ for any $S \in \mathcal{C}_i \setminus \{\{i\}, T\}$ (recall, that $u_{\{i\}} = 0$ for any $i \in N$). It is clear that $\Phi_p(u)_i < 0$, which contradicts $H(u) = C(\alpha_u) \subseteq \Re_+^N$.

On the other hand, due to the formula (2.1) for any $p \in \mathcal{P}$ and $S \subseteq N$ we get

$$\Phi_p(u)(S) = u(S) + \sum_{i \in S} \sum_{T \in \mathcal{C}_i(S)} p_i^T u_T$$

with $\mathcal{C}_i(S) := \{T \in \mathcal{C}_i \mid T \setminus S \neq \emptyset\}$. Hence, assuming $u \in V_+$, we have $\Phi_p(u) \in C(u)$ for any $p \in \mathcal{P}$. Therefore, by Theorem 2.2 we deduce the inclusion $H(u) \subseteq C(u)$. Combining this with the inclusion $C(u) \subseteq H(u)$ already proven, we indeed obtain $C(\alpha_u) = C(u) = H(u)$ for any $u \in V_+$. □
To give more subtle description of the cores in terms of H-payoffs it seems reasonable to isolate some special classes of cooperative games and corresponding subsets of the (universal) vector space $\mathcal{H}$. Below, an example of this type of analysis is given.

Recall from Aumann and Shapley (1974) that c.g. v is called monotone if $v(S) \leq v(T)$ whenever $S \subseteq T$. Denote by $\mathcal{H}_o$ a collection of linear operators Φ belonging to $\mathcal{H}$ and satisfying

$$\Phi(BV_+) \subseteq \Re^N_+,$$

where BV_+ is the cone of monotone c.g. v belonging to V. So, due to the obvious inclusion $V_+ \subseteq BV_+$, an operator in $\mathcal{H}_o$ satisfies a stronger positivity requirement than just assigning nonnegative payoffs to totally positive games. We now denote $H_o(v) := \{\Phi(v) \mid \Phi \in \mathcal{H}_o\}$ and demonstrate that $H_o(v)$ coincides with $C(\alpha_v)$ in case v is a convex game.

To start with, let us prove first a representation theorem for the operators $\Phi \in \mathcal{H}_o$. Denote by $\mathcal{P}_o$ a family of Harsanyi dividend systems $p = [p_i^T]_{i \in N}^{T \in \mathcal{C}_i} \in \mathcal{P}$, satisfying the inequalities

$$\sum_{T \subseteq N \setminus S} (-1)^{|T|} p_i^{S \cup T} \geq 0, \quad i \in S, \ \ S \in \mathcal{C} \setminus \{N\}. \tag{2.16}$$

Theorem 2.14 *An operator Φ belongs to $\mathcal{H}_o$ iff $\Phi = \Phi_p$ for some $p \in \mathcal{P}_o$.*

Proof. Fix $p \in \mathcal{P}$ and for any $S \in \mathcal{C}$ put

$$q_i^S(p) := \sum_{T \subseteq N \setminus S} (-1)^{|T|} p_i^{S \cup T}, \quad S \in \mathcal{C}_i, \ i \in N. \tag{2.17}$$

We show that for any $i \in N$ and $v \in V$ it holds that

$$\Phi_p(v)_i = \sum_{T \in \mathcal{C}_i} q_i^T(p)[v(T) - v(T \setminus i)]. \tag{2.18}$$

Indeed, applying equality (2.1) we obtain $v(T) - v(T \setminus i) = \sum_{R \in \mathcal{C}_i | R \subseteq T} v_R$. Therefore $\sum_{T \in \mathcal{C}_i} q_i^T(p)[v(T) - v(T \setminus i)] = \sum_{T \in \mathcal{C}_i} [\sum_{R | T \subseteq R} q_i^R(p)] v_T$. Since $\Phi_p(v)_i = \sum_{T \in \mathcal{C}_i} p_i^T v_T$, it is fairly well to verify that

$$\sum_{R|T \subseteq R} q_i^R(p) = p_i^T, \quad T \in \mathcal{C}_i, \ i \in N. \tag{2.19}$$

By definition of $q_i^R(p)$ we have

$$\sum_{R|T \subseteq R} q_i^R(p) = \sum_{R|T \subseteq R} [\sum_{S \subseteq N \setminus R} (-1)^{|S|} p_i^{R \cup S}].$$

Summing up the coefficients at $p_i^{S \cup T}$ (with $S \subseteq N \setminus T$), we obtain

$$\sum_{R|T \subseteq R} q_i^R(p) = \sum_{S \subseteq N \setminus T} [\sum_{R \subseteq S} (-1)^{|R|}] p_i^{S \cup T}.$$

Taking into account formula (2.10) one gets relation (2.19).

Let now $p \in \mathcal{P}_o$. Then, by definition of $\mathcal{P}_o$ we have: $q_i^T(p) \geq 0$ for any $T \in \mathcal{C}_i$, $i \in N$. Therefore, on the basis of (2.18), one get: $\Phi_p(v) \in \Re_+^N$ for any $v \in BV_+$. On the other hand, in case $\Phi \in \mathcal{H}_o$ there exists (by Theorem 2.2) some $p \in \mathcal{P}$ such that $\Phi = \Phi_p$. Due to the monotonicity of functions $\widehat{\epsilon}_T$, $T \in \mathcal{C} \setminus \{N\}$, where

$$\widehat{\epsilon}_T(S) := \begin{cases} 1, & \text{if } T \subset S, \\ 0, & \text{otherwise,} \end{cases}$$

vectors $\Phi(\widehat{\epsilon}_T)$ are nonnegative for any $T \in \mathcal{C} \setminus \{N\}$ (here and below $T \subset S$ means $T \subseteq S$ and $T \neq S$). Further, because of (2.18), for any $\emptyset \neq T \subset N$ the equalities $\Phi(\widehat{\epsilon}_T)_i = q_i^{T \cup i}(p)$, $i \in N \setminus T$, hold. From this we obtain $q_i^T(p) \geq 0$, $T \in \mathcal{C}_i$, $i \in N$, which completes the proof. □

Observe that from the considerations given in the proof of Theorem 2.14 it follows that for any $p \in \mathcal{P}_o$, there exist $q_i^T \geq 0$, satisfying

$$\sum_{i \in R} \sum_{T|R \subseteq T} q_i^T = 1, \quad R \in \mathcal{C}, \tag{2.20}$$

such that the following representation holds:

$$p_i^R = \sum_{T|R \subseteq T} q_i^T, \quad R \in \mathcal{C}_i, \ i \in N. \tag{2.21}$$

Since the unique solution of the linear system (2.21) is $q_i^T := q_i^T(p)$, $T \in \mathcal{C}_i$, $i \in N$, conversely the following assertion is valid: any nonnegative solution of the system (2.20) gives rise to some element $p \in \mathcal{P}_o$, calculated according to formula (2.21). This relation allows us to give an explicit representation of the elements of the polyhedron $\mathcal{P}_o$ and corresponding linear operators $\Phi \in \mathcal{H}_o$.

Below we also use another linear system, which is equivalent to (2.20).

Lemma 2.15 *Tuple $[q_i^T]_{i\in N}^{T\in\mathcal{C}_i}$ is a solution of the linear system (2.20) iff the following conditions are satisfied*

$$\sum_{i\in N} q_i^N = 1, \tag{2.22}$$

$$\sum_{i\in T} q_i^T = \sum_{j\in N\setminus T} q_j^{T\cup j}, \quad T \in \mathcal{C}\setminus\{N\}. \tag{2.23}$$

Proof. Follows by induction on $k = |N \setminus T|$ [1]. □

We can now give a characterization of the core of a convex cooperative game in terms of Harsanyi payoffs.

Theorem 2.16 *The core of a convex cooperative game v coincides with the set $H_o(v) := \{\Phi_p(v) \mid p \in \mathcal{P}_o\}$.*

Proof. Let v be a convex game. If x belongs to $C(\alpha_v)$, then by Theorem 2.5 there exist $t_\pi \geq 0$, $\pi \in \Pi$, such that $\sum_{\pi\in\Pi} t_\pi = 1$ and $x = \sum_{\pi\in\Pi} t_\pi v^\pi$. It is easy to verify that the linear operator $\Phi(u) := \sum_{\pi\in\Pi} t_\pi u^\pi$, $u \in V$, belongs to $\mathcal{H}_o$. According to Theorem 2.13 there exists $p_o \in \mathcal{P}_o$ such that $\Phi(u) = \Phi_{p_o}(u)$ for any $u \in V$. Specifically, $x = \Phi_{p_o}(v)$, which means that the core-imputation x belongs to $H_o(v)$.

As to the inverse inclusion, $H_o(v) \subseteq C(\alpha_v)$, it is sufficient to prove that for any $[q_i^T]_{i\in N}^{T\in\mathcal{C}_i} \geq 0$ satisfying condition (2.20) there exists a nonnegative solution of the linear system

$$\sum_{\pi \mid T_i^\pi = T} t_\pi = q_i^T, \quad T \in \mathcal{C}_i,\ i \in N \tag{2.24}$$

(here and below $T_i^\pi := \{i_1, \ldots, i_k\}$, $i_k := i$, and $i_1, \ldots, i_{k-1}$ are all the elements preceding i in the permutation $\pi = (i_1, \ldots, i_n)$). Indeed, let $x = \Phi_p(v)$ for some $p \in \mathcal{P}_o$. As it was already mentioned in the proof of Theorem

[1] More details may be found in Vasil'ev (2003).

2.14, vector $q(p)$, defined by (2.17), is nonnegative and meets all the requirements (2.20) and (2.21). Assuming system (2.24) to have nonnegative solution $(t_\pi)_{\pi\in\Pi}$ for the right-hand side, defined by the vector $q(p)$, we get: $\sum_{\pi\in\Pi} t_\pi = \sum_{i\in N} q_i^N(p) = 1$. Further, because of (2.19) and (2.24), with $q = q(p)$, we obtain: $\sum_{\pi|T\subseteq T_i^\pi} t_\pi = p_i^T$, $T \in \mathcal{C}_i$, $i \in N$. Hence, for any $i \in N$ it holds that

$$x_i = \sum_{T\in\mathcal{C}_i} (\sum_{\pi|T\subseteq T_i^\pi} t_\pi) v_T = \sum_{\pi\in\Pi} t_\pi (\sum_{T\in\mathcal{C}_i|T\subseteq T_i^\pi} v_T).$$

From here, taking into account the equalities $\sum_{T\in\mathcal{C}_i|T\subseteq T_i^\pi} v_T = v^\pi(T_i^\pi) - v^\pi(T_i^\pi \setminus i) = v_i^\pi$, we obtain

$$x_i = \sum_{\pi\in\Pi} t_\pi v_i^\pi, \quad i \in N. \tag{2.25}$$

Since c.g. v is convex, due to Theorem 2.5 vector v^π belongs to $C(\alpha_v)$ for any $\pi \in \Pi$. Consequently, vector x belongs to the convex set $C(\alpha_v)$ as well, being by (2.25) a convex combination of the vectors v^π.

So, to prove the theorem we need to show the existence of a nonnegative solution to the linear system (2.24) for nonnegative q satisfying (2.20). To construct such a solution we follow (almost) the same idea, as in Dubey *et al.* (1981). The main essential addition consists of a more careful consideration of the irregular case, when at least one component of the vector $q = [q_i^T]_{i\in N}^{T\in\mathcal{C}_i}$ is equal to zero. So, let us analyze first the regular case, when q satisfies (2.20) and $q_i^T > 0$ for any $T \in \mathcal{C}_i$ and $i \in N$. To do so, set $q(\emptyset) := 1$, $q(T) := \sum_{i\in T} q_i^T$, for any $T \in \mathcal{C}$, and for any $\pi \in \Pi$ define

$$t_\pi^o := \prod_{i\in N} q_i^{T_i^\pi} / q(T_i^\pi \setminus i).$$

From definition of t_π^o we obtain

$$\sum_{\pi|T_i^\pi = T} t_\pi^o = [q_i^T / q(T\setminus i)] a_i^T b^T, \tag{2.26}$$

where

$$a_i^T := \sum_{\pi\in\Pi(T\setminus i)} \prod_{j\in T\setminus i} q_j^{T_j^\pi} / q(T_j^\pi \setminus j),$$

$$b^T := \sum_{\pi\in\Pi(N\setminus T)} \prod_{j\in N\setminus T} q_j^{T\cup T_j^\pi} / q(T \cup (T_j^\pi \setminus j))$$

(with $b^N := 1$, and $a_i^{\{i\}} := 1$, for any $i \in N$). Taking into account the evident relations

$$a_i^T = \sum_{j \in T \setminus i} [q_j^{T \setminus i} / q(T \setminus \{i,j\})] a_j^{T \setminus i},$$

$$b^T := \sum_{j \in N \setminus T} [q_j^{T \cup j} / q(T)] b^{T \cup j},$$

and applying induction on the number of elements in T and $N \setminus T$, respectively, we obtain (due to Lemma 2.15 for b^T): $a_i^T = q(T \setminus i)$, and $b^T = 1$ for any $T \in \mathcal{C}$. Thus, because of (2.26) we get: vector $(t_\pi^o)_\Pi$ is a nonnegative solution of the system (2.24), and, hence, we have proved our assertion for the regular case.

As to the irregular case, together with $q = [q_i^T] = [q_i^T]_{i \in N}^{T \in \mathcal{C}_i}$, satisfying condition (2.20), we consider the sequence $(q_m)_{m=1}^\infty = ([q_{i,m}^T])_{m=1}^\infty$ with $q_{i,m}^T := (1 - \frac{1}{m}) q_i^T + \frac{1}{m} \bar{q}_i^T$, and $\bar{q}_i^T := \frac{(n-|T|)!(|T|-1)!}{n!}$ for any $T \in \mathcal{C}_i$, $i \in N$ (i.e., taking $\bar{q}_i^T$ to be the well-known Shapley shares). It is easy to verify that (i) $q_m \gg 0$, (ii) q_m satisfies requirement (2.20) for any $m \geq 1$, and (iii) $\lim_{m \to \infty} q_m = q$. From (i), (ii), and analysis given above for the regular case, it follows immediately that for any $m \geq 1$ there exists a nonnegative solution $t^m = (t_\pi^m)_{\pi \in \Pi}$ of the linear system (2.24) with the right hand sides equal to $q_{i,m}^T$, $T \in \mathcal{C}_i$, $i \in N$. It is clear that sequence $(t^m)_{m=1}^\infty$ is bounded and, hence, contains some convergent subsequence. Due to the property (iii) of the sequence $(q_m)_{m=1}^\infty$, the limit $(t_\pi)_\Pi$ of the convergent subsequence, mentioned above, satisfies all the equalities (2.24). □

Remark 2.17 Let us mention first, that the vector $[\bar{q}_i^T]_{i \in N}^{T \in \mathcal{C}_i}$, we used above for regularization of (irregular) q, corresponds to the Shapley value dividends $\bar{p}_i^T = \frac{1}{|T|}$, $T \in \mathcal{C}_i$, $i \in N$, in the sense that $\bar{q}$ satisfies equalities

$$\bar{p}_i^T = \sum_{R | T \subseteq R} \bar{q}_i^R, \quad T \in \mathcal{C}_i,\ i \in N.$$

Further, the way in which we have proved that the linear system (2.24) has a nonnegative solution, can be used to prove that each extreme point q^* of the convex polyhedron $\mathcal{Q}$, being a nonnegative solution of the linear system (2.20), is of the form $q^* = q^\pi$, for some $\pi \in \Pi$, where

$$(q^\pi)_i^T = q_i^{\pi,T} := \begin{cases} 1, & \text{if } T = T_i^\pi, \\ 0, & \text{otherwise.} \end{cases}$$

Actually, what is left to complete this proof consists of the rather simple verification of the inclusions: $q^\pi \in \operatorname{ex} \mathcal{Q}$ for any $\pi \in \Pi$ (for more details see Vasil'ev, 2003).

To conclude this section, we give a generalization of Theorem 2.13 for the games which may not satisfy condition (2.6). We first mention that a characterization of the core of some cooperative game v in terms of H-payoffs makes sense only provided that the set $H^+(v)$ of individually-rational H-payoffs (the Harsanyi Imputation set of v), defined by the formula $H^+(v) := H(v) \cap I(v)$, is nonempty. From Vasil'ev (1981a) it is known that $H^+(v) \neq \emptyset$ iff

$$v(N) \geq v_H(S) + \sum_{N \setminus S} v(i), \quad S \subseteq N. \tag{2.27}$$

Below, it will be shown that this minimal condition yields the inclusion $C(\alpha_v) \subseteq H^+(v)$.

To be more precise, the next theorem states that $H^+(v)$ is external stable (in the sense of von Neumann and Morgenstern, 1953), when not empty. Recall that for an imputation $x \in I(v)$, we say that $y \in I(v)$ dominates x, denoted by $x \; \alpha_v \; y$, when there exists a coalition S such that $y_i > x_i$ for all $i \in S$ and $y(S) = \sum_{i \in S} y_i \leq v(S)$. A set $Y \subseteq I(v)$ is external stable when for every $x \in I(v) \setminus Y$ there exists $y \in Y$ such that y dominates x.

Theorem 2.18 *Assume v satisfies condition (2.27). Then $H^+(v)$ is external stable.*

Proof. As it was already mentioned, $H^+(v) \neq \emptyset$, provided that v satisfies condition (2.27). Hence, under the assumption of our theorem there exists at least one element of $H(v)$ belonging to $I(v)$. Further, since both $I(v) = I(\widehat{v}) + v_{(1)}$ and $H^+(v) = H^+(\widehat{v}) + v_{(1)}$ with $\widehat{v} := v - v_{(1)}$, without loss of generality we may assume that v is 0-normalized, i.e. $v(i) = 0$, for all $i \in N$, and thus $\widehat{v}(S) = v(S)$ for all $S \subseteq N$. Denote by $\bar{v}_H$ the function defined by

$$\bar{v}_H(S) := \max\{v_H(T) \mid T \subseteq S\}, \quad S \subseteq N.$$

It is not very hard to verify that $\bar{v}_H$ is a convex function and that $H^+(v) = C(\bar{v}_H)$ (for more details see, e.g. Vasil'ev, 1981a).

It is clear that when $I(v) \setminus H^+(v) = \emptyset$, the theorem is true. So, consider the case $I(v) \setminus H^+(v) \neq \emptyset$ and let x be any element belonging to $I(v) \setminus H^+(v)$. Since $x \notin H^+(v) = C(\bar{v}_H)$, there exists $S \subseteq N$ such that $x(S) < \bar{v}_H(S)$.

Let S_0 be a minimal coalition satisfying this inequality, i.e., $x(S_0) < \bar{v}_H(S_0)$ and $x(T) \geq \bar{v}_H(T)$ for any $T \subseteq S_0$, $T \neq S_0$. Define $x^0 = (x_i^0)_{i \in S_0}$ by

$$x_i^0 := x_i + \delta, \quad i \in S_0,$$

where $\delta = \frac{1}{|S_0|}[\bar{v}_H(S_0) - x(S_0)]$. Clearly $x^0(S_0) = \sum_{i \in S_0} x_i^0 = \bar{v}_H(S_0)$ and thus $x^0 \in C(\bar{v}_H^0)$, where $\bar{v}_H^0$ is the restriction of $\bar{v}_H$ to S_0. Due to the already mentioned convexity of the game $\bar{v}_H$, there exists an extension $y \in \Re^N$ of x^0 in the c-core of the game $\bar{v}_H$, i.e., a vector y such that $y_i = x_i^0$ for $i \in S_0$ and $y \in C(\bar{v}_H)$. Thus, by virtue of the equality $C(\bar{v}_H) = H^+(v)$, payoff vector y belongs to $H^+(v)$ and satisfies $y_i = x_i^0 > x_i$, $i \in S_0$ and $y(S_0) = \bar{v}_H(S_0)$. To show that y dominates x through S_0 in the game v it remains to prove that $\bar{v}_H(S_0) \leq v(S_0)$. Since the negative variation v^- of v is convex and thus superadditive, we have by definition of v_H that $v_H(S_0) \leq v(S_0)$. Now, suppose that $\bar{v}_H(S_0) > v_H(S_0)$. Then, by definition of $\bar{v}_H$, there exists $T \subseteq S_0$, $T \neq S_0$, such that $\bar{v}_H(S_0) = v_H(T)$. Since x is nonnegative, we have that $x(T) \leq x(S_0) < \bar{v}_H(S_0) = v_H(T) \leq \bar{v}_H(T)$, which contradicts the inequality $x(T) \geq \bar{v}_H(T)$. Hence $\bar{v}_H(S_0) = v_H(S_0) \leq v(S_0)$. □

Recall once more from von Neumann and Morgenstern (1953) that the classic core of a game v is the set $C(\alpha_v)$ of undominated imputations, i.e.,

$$C(\alpha_v) := \{x \in I(v) \mid \text{ there is no } y \in I(v) \text{ such that } y \text{ dominates } x\}.$$

Then it follows from the external stability of the set $H^+(v)$ that the classic core is contained in the Harsanyi set if the latter set is not empty. So, we have the following corollary.

Corollary 2.19 *For any $v \in V$ with $H^+(v) \neq \emptyset$ it holds $C(\alpha_v) \subseteq H^+(v)$.*

Remark 2.20 Theorem 2.18 demonstrates that the Harsanyi Imputation set $H^+(v)$ (individually-rational H-payoffs) of a c.g. v, satisfying the requirement (2.27)), dominates "as a whole" the other (classic) individually rational imputations of v. The principle of sequential partitioning of the partially ordered sets under consideration over two subsets, dominated and dominating components, is the basis for the main concept in the next part of this paper, namely the concept of a generalized von Neumann-Morgenstern solution of a cooperative game.

3 gNM-Solutions and accessibility of cores

This section deals with the external (outer) stability of some solutions of so-called abstract games (shortly a.g.), see Roth (1976). Solutions under consideration are mostly either the cores of abstract games (as before, we mainly pay attention to TU-cooperative games), or rather natural internally stable extensions of these cores, like classic NM-solutions and its generalizations. We study existence conditions for such extensions and analyze stability of the corresponding adjustment processes of sequential improvement of dominated alternatives, i.e., we investigate the hypothesis' for monotone trajectories to converge to the solutions in question. On the whole, the problems we consider below are closely related to the bargaining stabilization conditions, especially to the questions of the efficient formation of blocking coalitions and dominating imputations.

In more detail, we deal with some generalizations of the von Neumann-Morgenstern solution (1953). The main concept of generalized NM-solution (shortly gNM-solution) is based on the principle of sequential (dynamic) improvement of dominated alternatives. The notion of NM-rank, characterizing the measure of irregularity of games without classic NM-solution, is introduced, and some gNM-existence theorems with the evaluation of NM-rank are established. We also present a definition of the limit transitive closure of the standard domination relation in cooperative games by allowing infinite convergent monotone trajectories. One of the main results of this paper obtained within the framework of the theory of dynamic systems (Maschler and Peleg, 1976), consists of the external stability of the cores of balanced TU-games w.r.t. the above mentioned limit transitive closure of the classic domination relation α_v. In more detail, we prove the *accessibility* of the core of any balanced TU-cooperative game v : for any $x \in I(v) \setminus C(\alpha_v)$, there exists a convergent trajectory $x_0, x_1, \ldots$ such that $x_0 = x$, $\lim x_r \in C(\alpha_v)$ and $x_r \; \alpha_v \; x_{r+1}$ for any $r = 0, 1, \ldots$. To show that the TU-property is essential we construct an example of an NTU-game with a "black hole", being a closed subset of imputations that doesn't intersect the corresponding core and contains all the monotone trajectories originating from this subset.

3.1 gNM-Solutions: basic definitions and notations

The basic definitions and constructions introduced below are given in terms of the above mentioned abstract games (Roth, 1976), allowing to apply them for the analysis of a wide class of solutions, based on the binary domination relations.

Definition 3.1 Let X be some set, and $\alpha \subseteq X \times X$ an arbitrary nontrivial binary relation on X. A system $\Gamma = (X, \alpha)$ is called an abstract game (a.g.).

According to Vasil'ev (1984) a binary relation $\alpha \subseteq X \times X$ is called nontrivial, if there exist $x, y \in X$ such that $(x,y) \in \alpha$, and $x \neq y$ (i.e., α is nonempty and contains at least one pair (x,y) with $x \neq y$.) As usual (von Neumann and Morgenstern, 1953), we say that a set $Y \subseteq X$ is internally stable (w.r.t. α) if for any two distinct elements $x, y \in Y$ the pair (x,y) doesn't belong to α. Recall also (see von Neumann and Morgenstern, 1953) that a set $Y \subseteq X$ is said to be externally stable (w.r.t. α), if for any $x \in X \setminus Y$ there is $y \in Y$ such that $(x,y) \in \alpha$ (to ease the interpretation, recall when $(x,y) \in \alpha$ we treat "y as at least as good as x").

Denote by α_* the transitive closure of α : $(x,y) \in \alpha_*$ iff there exists a family $\{x_0, x_1, \ldots, x_m\}$ with $x_0 = x$, $x_m = y$ satisfying the following conditions (below we use the standard shortening: $x \ \beta \ y \Leftrightarrow (x,y) \in \beta$, for any $\beta \subseteq X \times X$)

$$x_r \ \alpha \ x_{r+1}, \quad \text{for any } r = 0, 1, \ldots, m-1. \tag{3.1}$$

Definition 3.2 Let $\Gamma = (X, \alpha)$ be an a.g. A set $Y \subseteq X$ is called weakly externally stable, if for any $x \in X \setminus Y$ there is $y \in Y$ such that $x \ \alpha_* \ y$.

Recall from von Neumann and Morgenstern (1953) that an internally stable set $Y \subseteq X$ of an a.g. $\Gamma = (X, \alpha)$ is said to be a von Neumann-Morgenstern solution (shortly NM-solution) of Γ, if it is externally stable (w.r.t. α).

One of the generalizations of the classic NM-solution rests on the principle of finite sequential (dynamic) improvement of dominated alternatives.

Definition 3.3 An internally stable set $Y \subseteq X$ of an a.g. $\Gamma = (X, \alpha)$ is called a generalized NM-solution (gNM-solution) of Γ, if this set is weakly externally stable.

Thus, a set $Y \subseteq X$ is a generalized NM-solution of $\Gamma = (X, \alpha)$, if this set is internally stable w.r.t. α, and externally stable w.r.t. α_*. Consequently, in case the binary relation α of an a.g. $\Gamma = (X, \alpha)$ is transitive, any gNM-solution of the game Γ is a (classic) NM-solution of that game.

For any $(x,y) \in \alpha_*$ denote by $m(x,y)$ the smallest of natural numbers m, for which the family $\{x_0, x_1, \ldots, x_m\}$ satisfying monotonicity condition (3.1) with $x_0 = x$, $x_m = y$, exists. Let Y be some generalized NM-solution of an a.g. $\Gamma = (X, \alpha)$. For $x \in X \setminus Y$ we denote by $m(x)$ the smallest of the numbers $m(x,y)$ with $y \in Y$ satisfying condition $x \ \alpha_* \ y$, i.e.,

$$m(x) := \min\{m(x,y) \mid y \in Y \ \text{ such that } \ x \ \alpha_* \ y\}$$

(as usual we define $m(x) := +\infty$) in case $(x, y) \notin \alpha_*$ for all $y \in Y$). We say that a gNM-solution Y is *finitary* , if the magnitude

$$r(Y) = \sup\{m(x) \mid x \in X \setminus Y\}, \tag{3.2}$$

to be called the *NM-rank of* Y, is finite (otherwise we put $r(Y) := +\infty$). To characterize the measure of irregularity of games without NM-solutions we introduce the notion of the NM-rank of an a.g.

Definition 3.4 Let an a.g. Γ admits at least one finitary generalized NM-solution. Then the magnitude

$$r(\Gamma) = \min\{r(Y) \mid Y \text{ is a generalized NM-solution of } \Gamma\} \tag{3.3}$$

is called the NM-rank of Γ.

Observe that from the definition of NM-rank it follows that $r(\Gamma) = 1$ if Γ has a (classic) NM-solution. Note also that the NM-rank of the famous Lucas 10-person TU-cooperative game, which has no NM-solution, is equal to two (Vasil'ev, 1981b). Some examples of abstract games of any preassigned rank greater than one are given in Vasil'ev (1984) (see, also, Makarov and Vasli'ev, 1989).

To study monotone trajectories corresponding to the cooperative games, we need some general results about generalized NM-solutions of an a.g. (X, α) in case that X is provided with some topology τ. One useful generalization of the weak external stability in the presence of τ is as follows:

(L) Any τ-neighborhood of Y is weakly externally stable.

Definition 3.5 For some given topology τ, an internally stable set Y is called a limiting generalized NM-solution of an a.g. (X, α) if it satisfies condition (L).

3.2 NM-Rank of cooperative games

The type of generalized NM-solution of an a.g. $\Gamma = (X, \alpha)$ depends on the topological and order characteristics of the upper and lower sections $\alpha(x) := \{z \mid (x, z) \in \alpha\}$ and $\alpha^{-1}(x) := \{y \mid (y, x) \in \alpha\}$, respectively, as well as on the structure of the core of Γ, defined in terms of the upper sections $\alpha(x)$ of the binary relation α. Recall that the core is the set of all the α-maximal elements of X, i.e.,

$$C(\alpha) := \{x \in X | \alpha(x) = \emptyset\}.$$

The core is contained in any generalized NM-solution of Γ (if such exists). Moreover, under some natural assumptions $C(\alpha)$ is a limiting generalized NM-solution of Γ. To give an example, recall from von Neumann and Morgenstern (1953) that a binary relation β is called acyclic, if its transitive closure β_* is irreflexive (i.e., $(x,x) \notin \beta_*$ for any $x \in X$).

Theorem 3.6 *Let $C(\alpha) \neq \emptyset$. If X is compact and there exists an acyclic restriction $\beta \subseteq \alpha$, for which $C(\beta) = C(\alpha)$, and the lower sections $\beta^{-1}(x)$ are open for all $x \in X$, then $C(\alpha)$ is a limiting generalized NM-solution of the a.g. $\Gamma = (X, \alpha)$.*

Proof. Let U be an arbitrary τ-neighborhood of $C(\alpha)$. For every $x \notin U$ fix some $z(x) \in \beta(x)$ and put $U_x = \beta^{-1}(z(x))$. It is clear that $\{U_x\}_{x \in X \setminus U}$ is an open cover of $X \setminus U$. Since $X \setminus U$ is a compact set, there exists a finite subset $Z \subseteq \{z(x) \mid x \in X \setminus U\}$ such that

$$X \setminus U \subseteq \bigcup_{z \in Z} \beta^{-1}(z).$$

Let x be an arbitrary element of $X \setminus U$. Put $x_0 = x$ and choose x_1 to be any element of Z satisfying $x_0\ \beta\ x_1$. If $x_1 \notin U$ we proceed by picking some $x_2 \in Z$ satisfying $x_1\ \beta\ x_2$ and so on. Since β is acyclic and Z is finite, the above-mentioned process results in the formation of the family $\{x_0, x_1, \ldots, x_m\}$ with $x_m \in U$ for some $m = m_x$ and $x_{r-1}\ \beta\ x_r$ for all $r = 1, \ldots, m$. Hence, putting $y = x_m$, by the inclusion $\beta \subseteq \alpha$ we obtain that $x\ \alpha_*\ y$ with y belonging to U. □

This result is supplemented by

Theorem 3.7 *Let $\Gamma = (X, \alpha)$ be an a.g. with X a compact set. If there exists a restriction $\beta \subseteq \alpha$, for which $C(\beta) = \emptyset$, and the lower sections $\beta^{-1}(x)$ are open for every $x \in X$, then Γ has a finitary generalized NM-solution Y such that $|Y| < \infty$ and $r(Y) \leq 3$.*

For the sake of completeness, we give first a proof of the following auxiliary (folk) lemma (Vasil'ev, 1984), which may be of interest in itself. Namely, this lemma demonstrates that there always exists a gNM-solution of a finite game Γ. Moreover, any finite game $\Gamma = (X, \alpha)$ has at least one gNM-solution Y with NM-rank $r(Y) \leq 2$. Recall, that we call $\Gamma = (X, \alpha)$ a finite abstract game, if $|X| < \infty$.

Lemma 3.8 *The NM-rank of any finite a.g. $\Gamma = (X, \alpha)$ is less or equal to 2.*

Proof. As in Vasil'ev (1984), we apply induction on $|X|$. It is clear, that our assertion is true for $|X| \leq 3$. Suppose, it is valid for any finite a.g. (X, α) with $|X| \leq m$, and consider an arbitrary a.g. $\Gamma = (Z, \beta)$ with $|Z| = m + 1$. Since β is nontrivial, there exists a pair $(x, y) \in \beta$ such that $x \neq y$. Put $\widehat{Z} := Z \setminus (\beta^{-1}(y) \cup \{y\})$, and consider two possibilities: 1) $\widehat{Z} = \emptyset$ and 2) $\widehat{Z} \neq \emptyset$. Obviously, in the first situation we have $r(\{y\}) = 1$. As to the second possibility, denote by $\widehat{\beta}$ the restriction of β to $\widehat{Z}$ (i.e., take $\widehat{\beta} := \beta \bigcap \widehat{Z} \times \widehat{Z}$) and observe, that when $\widehat{\beta}$ is trivial, we get $r(\Gamma) \leq 2$. Indeed, putting $Y := \widehat{Z}$ in case $\beta(y) \bigcap \widehat{Z} \neq \emptyset$, and $Y := \widehat{Z} \cup \{y\}$ otherwise, we obtain in both these cases that Y is a gNM-solution of Γ with $r(Y) \leq 2$. Finally, letting $\widehat{\beta}$ to be nontrivial, by the inductive hypothesis we have: there exists a gNM-solution $\widehat{Y}$ of the a.g. $\widehat{\Gamma} = (\widehat{Z}, \widehat{\beta})$ such that $r(\widehat{Y}) \leq 2$. Again, by putting $Y := \widehat{Y}$ in case $\beta(y) \bigcap \widehat{Y} \neq \emptyset$, and $Y := \widehat{Y} \cup \{y\}$ otherwise, we have that in both these cases Y is a gNM-solution of Γ with $r(Y) \leq 2$. □

Proof of Theorem 3.7. Since $C(\beta) = \emptyset$, the family $\{\beta^{-1}(x)\}_{x \in X}$ is an open cover of X. By compactness of X there is a finite subset $\widehat{Z} \subseteq X$ such that $X = \bigcup_{z \in \widehat{Z}} \beta^{-1}(z)$. Denote by $\widehat{\alpha}$ the restriction of α to $\widehat{Z}$. It is clear, that in case $\widehat{\alpha}$ is trivial we have: $\widehat{Z}$ is a finite gNM-solution of Γ with $r(\widehat{Z}) = 1$. In case $\widehat{\alpha}$ is non-trivial, put $\widehat{\Gamma} := (\widehat{Z}, \widehat{\alpha})$. From Lemma 3.8 it follows that $r(\widehat{\Gamma}) \leq 2$. Let $\widehat{Y}$ be a gNM-solution of $\widehat{\Gamma}$ with $r(\widehat{Y}) \leq 2$. Put $\widehat{Z}_1 = \widehat{Y}$, and define $\widehat{Z}_2$, $\widehat{Z}_3$ by

$$\widehat{Z}_2 = \{z \in \widehat{Z} \mid \exists y \in \widehat{Z}_1 \text{ such that } z \,\widehat{\alpha}\, y\},$$

$$\widehat{Z}_3 = \{z \in \widehat{Z} \setminus (\widehat{Z}_1 \cup \widehat{Z}_2) \mid \exists y \in \widehat{Z}_2 \text{ such that } z \,\widehat{\alpha}\, y\}.$$

Since $r(\widehat{Y}) \leq 2$, it follows that the collection $\{\widehat{Z}_1, \widehat{Z}_2, \widehat{Z}_3\}$ is a partition of $\widehat{Z}$. Further, by construction we have that $\widehat{Z}_3 \,\widehat{\alpha}\, \widehat{Z}_2$, and $\widehat{Z}_2 \,\widehat{\alpha}\, \widehat{Z}_1$, where

$$B \,\widehat{\alpha}\, D \Leftrightarrow \forall b \in B \text{ there exists } d \in D \text{ such that } b \,\widehat{\alpha}\, d \tag{3.4}$$

(here $\emptyset \,\widehat{\alpha}\, D$ for any D by definition). Hence, from $(X \setminus \widehat{Z}) \,\beta\, \widehat{Z}$ and $\beta \subseteq \alpha$ it follows that $m(x) \leq 3$ for any $x \in X \setminus \widehat{Y}$. Therefore NM-rank $r(\widehat{Y}) = \sup \{m(x) \mid x \in X \setminus \widehat{Y}\}$ of the internally stable (w.r.t. α) finite set $\widehat{Y}$ in (X, α) is at most three and, consequently, $\widehat{Y}$ is a finitary gNM-solution of the a.g. $\Gamma = (X, \alpha)$. □

Since α_* is transitive for any binary relation α, and $\alpha_*^{-1}(x)$ is open for any $x \in X$ whenever $\alpha^{-1}(x)$ is open for any $x \in X$, we get the following consequence of Theorem 3.7.

Corollary 3.9 *If X is a compact set, $C(\alpha) = \emptyset$, and lower sections $\alpha^{-1}(x)$ are open for every $x \in X$, then the a.g. (X, α_*) has a finite (classic) NM-solution.*

It is known (Vasil'ev, 1981b), that the NM-rank of a TU-cooperative game is less or equal to $2^n - n - 2$, where n is the number of players. Theorem 3.7 brings the following refinement of this result.

Corollary 3.10 *The NM-rank of any TU-cooperative game with empty core is less or equal to 3.*

We will point out also a specification of the general results obtained for the rather wide class of NTU-cooperative games Γ_G, given in the form of characteristic functions G:

$$S \longmapsto G(S), \quad S \in \bar{\Omega},$$

where $\bar{\Omega} = \Omega \cup \{N\}$, $\Omega \subseteq \mathcal{C} = 2^N \setminus \{\emptyset\}$ is a nonempty family of admissible blocking coalitions, $N = \{1, \ldots, n\}$, as before, is a set of players, and $G(S) \subseteq \Re^S$ is a set of imputations accessible through the efforts of the coalition S. The dominance relation α_G on $X_G = G(N)$ has the following standard form (below, y_S is the restriction of y to S):

$$x \; \alpha_G \; y \Leftrightarrow \exists S \in \Omega \text{ satisfying } [\forall i \in S (x_i < y_i) \text{ and } (y_S \in G(S))].$$

We stress that we do not assume (in contrast to the ordinary NTU-case) that $\Omega = \mathcal{C}$, and $G(S) - \Re^S_+ \subseteq G(S)$, $S \in \Omega$. Nevertheless, to mention one reduced form of the "free-disposal" assumption, we will say that a game $\Gamma_G = (X_G, \alpha_G)$ is satiated, if for any $x \in G(N)$, $S \in \Omega$, and $y \in G(S)$ it holds: $y \leq x_S$ implies $(y, x_{N \setminus S}) \in G(N)$. Recall also, that the weak Pareto boundary $\Theta = \Theta(G)$ of the set $G(N)$ is defined by

$$\Theta = \{x \in G(N) \mid y \gg x \Rightarrow y \notin G(N)\},$$

where, as usual, $y \gg x \Leftrightarrow \forall i \in N \; (y_i > x_i)$.

Theorem 3.11 *Let $G(S)$ be closed for every $S \in \bar{\Omega}$ and let $G(N)$ be bounded above. Then $r(\Gamma_G) \leq |\Omega|$ and, (unlike the classic NM-solution (Lukas, 1971)) the common part of all finitary generalized NM-solutions of the a.g. $\Gamma_G = (X_G, \alpha_G)$ coincides with $C(\alpha_G)$.*
If, moreover, Γ_G is satiated, $N \in \Omega$, the weak Pareto boundary Θ of $G(N)$ is bounded from below and $C(\alpha_G) = \emptyset$, then $r(\Gamma_G) \leq 3$.

Proof. Let us fix some enumeration of the blocking coalitions: $\Omega = \{S_1, \ldots, S_m\}$, and put

$$X_1 := \{x \in X_G \mid \exists y \in X_G \text{ such that } x \; \alpha_1 \; y\},$$

$$X_{r+1} := \{x \in X_G \setminus \bigcup_{s=1}^{r} X_s \mid \exists y \in X_G \setminus \bigcup_{s=1}^{r} X_s \text{ such that } x \; \alpha_{r+1} \; y\}, \; r = 1, \ldots, m-1,$$

$$X_{m+1} := X_G \setminus \bigcup_{s=1}^{m} X_s,$$

where

$$x \; \alpha_r \; y \Leftrightarrow [(y_{S_r} \in G(S_r)) \text{ and } (x_i < y_i, \; i \in S_r)].$$

Assume, for simplicity, that all the sets $X_1, \ldots, X_m$ are nonempty. Put $Y_0 := X_G, \; Y_r := X_G \setminus \bigcup_{s=1}^{r} X_s, \; r = 1, \ldots, m$. We prove that $X_r \; \alpha_r \; Y_r$ for all $r = 1, \ldots, m$ (here we use the notation (3.4)). First, for any $r = 1, \ldots, m$, the set X_r is (relatively) open in Y_{r-1}. Hence, due to the closedness and boundedness from above of the set X_G we have that Y_r is closed and bounded from above for any $r = 0, 1, \ldots, m$. Therefore, for every $r = 1, \ldots, m$ and $x \in X_r$ there exists $\varepsilon > 0$ such that the subset

$$Y_r(x) := \{y \in Y_{r-1} \mid x \; \alpha_r \; y \text{ and } \; y_i \geq x_i + \varepsilon, \text{ each } \; i \in S_r\}$$

is nonempty, closed, and bounded above. Put

$$B_r(x) := \{y \in Y_r(x) \mid y(S_r) \geq z(S_r) \text{ for any } z \in Y_r(x)\},$$

where, as before, $z(S) = \sum_{i \in S} z_i$ for any $z \in \Re^N, S \subseteq N$. It is clear, that $B_r(x) \neq \emptyset$, and $\{x\} \; \alpha_r \; B_r(x)$. Besides, $B_r(x) \subseteq Y_r$ (since otherwise we can choose $y \in B_r(x)$ and $z \in Y_{r-1}$ such that $y \; \alpha_r \; z$ and, consequently, $z \in Y_r(x)$, implying, together with the inequality $z(S_r) > y(S_r)$, a contradiction with the definition of $B_r(x)$). Thus, $Y_m = X_{m+1} \neq \emptyset$, and $Y_r \neq \emptyset, \; X_r \; \alpha_r \; Y_r$ for every $r = 1, \ldots, m$.

From the definitions of α_G and X_r it follows that X_{m+1} is internally stable w.r.t. α_G and $r(X_{m+1}) \leq m$. To complete the proof of the first part of the theorem, note that any coalition $S \in \Omega$ may be taken as S_1 under appropriate enumeration of Ω. Hence, for any $x \notin C(\alpha_G)$ there exists a finitary generalized NM-solution X_{m+1} of Γ_G such that $x \notin X_{m+1}$.

To prove the second part of the theorem, consider an auxiliary game Γ_F, defined by the characteristic function F: $F(N) = \Theta, \; F(S) = G(S)$, for any $S \in \Omega$, $S \neq N$. Recall, that Θ is the weak Pareto boundary of $G(N)$, defined by the formula $\Theta = \{x \in G(N) \mid y \gg x \Rightarrow y \notin G(N)\}$. Directly from the satiation property and definition of Θ it follows that if $x \in \Theta$ is dominated by an element of $G(N)$, it is dominated by some element of Θ, as well. Indeed, let, for example, $x \in \Theta$ is dominated by $x' \in G(N)$ via some coalition $S \in \Omega$. Put $Z := \{y' \in G(N) | x' \leq y' \text{ and } y'_S \in G(S)\}$. Due to the compactness of Z there exists $y \in Z$ with $y(N) = \max\{z(N) | z \in Z\}$. In case y does not belong to Θ, we can choose $\bar{y} \in G(N)$ such that $\bar{y}_i > y_i$ for every $i \in N$. It is clear, that due to the inequality $\bar{y}(N) > y(N)$ we have: $S \neq N$. Further, since Γ_G is satiated, we have: $(y_S, \bar{y}_{N \setminus S}) \in Z$, which contradicts (because of $N \setminus S \neq \emptyset$) the definition of y. Thus, $y \in \Theta$ and $x \; \alpha_G \; y$. Since $C(\alpha_G) = \emptyset$, the above-mentioned fact implies $C(\alpha_F) = \emptyset$. Further, due to the boundedness and closedness of Θ we have that Γ_F satisfies all the requirements of Theorem 3.7. To complete the proof, it remains to note that any gNM-solution of Γ_F is a gNM-solution of Γ_G with the same NM-rank (the latter follows from $N \in \Omega$ and $C(\alpha_F) = \emptyset$). □

3.3 Weak* external stability: a general approach

To introduce one more extension of the initial binary relation α on X, we recall that the core of TU-cooperative game, normally, is neither externally, nor weakly externally stable in $I(v)$ (see, e.g., Bondareva, 1962, and Kulakovskaya, 1971). In the sequel we assume that X is provided with some metrizable topology τ.

Definition 3.12 Let α be a binary relation on X. Then the binary relation $\alpha^* \subseteq X \times X$, defined by the formula

$$x \; \alpha^* \; y \;\; \Leftrightarrow \;\; \exists \{x_r\}_{r=0}^{\infty} \subseteq X [x_0 = x, \; \lim_{r \to \infty} x_r = y, \\ \text{and } x_r \; \alpha \; x_{r+1} \text{ for any } r = 0, 1, \ldots],$$

is the limit transitive closure of α.

We say that a sequence $\{x_r\}_{r=0}^{\infty}$ is *α-monotone* (or shortly *monotone*), if $x_r \; \alpha \; x_{r+1}$ for any $r = 0, 1, \ldots$. Thus, $x \; \alpha^* \; y$ means that there exists a convergent α-monotone sequence, starting from x, whose limit equals y. Observe also, that directly from definition 3.12 it follows: α^* is transitive and $\alpha^{*-1}(x) = (\alpha^*)^{-1}(x)$ is open for any $x \in X$ whenever $\alpha^{-1}(x)$ is open for any $x \in X$. Moreover, putting $(x, y)_\alpha := \alpha(x) \cap \alpha^{-1}(y)$, we have that in case $(x, y)_\alpha \bigcap U_y \neq \emptyset$, for each pair $(x, y) \in \alpha$ and each neighborhood U_y of y, the limit transitive closure α^* is an extension of α, i.e., $\alpha \subseteq \alpha^*$.

Hence, in case $\alpha^{-1}(x)$ is open for each $x \in X$, and an "interval" $(x, y)_\alpha$ contains elements, arbitrarily close to y, for any $(x, y) \in \alpha$, we obtain: α^* is a transitive extension of α with ${\alpha^*}^{-1}(x)$ to be open for any $x \in X$.

Definition 3.13 Let $\Gamma = (X, \alpha)$ be an a.g. A set $Y \subseteq X$ is called weak* externally stable, if for any $x \in X \setminus Y$ there is a $y \in Y$ such that $x \; \alpha^* \; y$.

Remark 3.14 Recall from Vasil'ev (1987) that in case $Y = C(\alpha)$ is weak* externally stable, we say that the core $C(\alpha)$ of an a.g. $\Gamma = (X, \alpha)$ is accessible (c.f. Makarov and Vasli'ev, 1989, and Vasil'ev, 1984).

One of the main results of this paper, obtained within the framework of the theory of dynamic systems, demonstrates, among other things, that the core of any balanced TU-cooperative game v is accessible (i.e., $C(\alpha_v)$ is externally stable w.r.t. limit transitive closure of the classic domination relation α_v for any balanced TU-game v).

Apart from the natural connection with the questions considered earlier, the problem of establishing the accessibility of the cores of various games presents, actually, considerable interest of its own (see, e.g., Green, 1974; Maschler and Peleg, 1976). Some general results on weak* external stability of $C(\alpha)$ are set forth below, based on the study of suitable dynamic systems associated with the binary relation α.

In the following it is assumed that X is a complete metric space with metric d. By point-set dynamic systems (p.s.d.s.) on X, like in Maschler and Peleg (1976), we mean correspondences $\varphi : X \to 2^X$ for which $\varphi(x) \neq \emptyset$ for all $x \in X$. By (infinite) trajectories of φ we mean sequences $\{x_r\}_{r=0}^{\infty}$ for which $x_r \in \varphi(x_{r-1})$, $r = 1, 2, \ldots$.

For p.s.d.s. $\psi : X \to 2^X$ let $gr\,\psi := \{(x, y) \in X \times X \mid y \in \psi(x)\}$ be a graph of ψ and $\psi_* := \bigcup_{m=1}^{\infty} \psi^m$ be a p.s.d.s. ("expansion of ψ"), defined by the formula

$$\psi_*(x) := \bigcup_{m=1}^{\infty} \psi^m(x), \quad x \in X,$$

where $\psi^1(x) := \psi(x)$ and $\psi^{m+1}(x) := \bigcup_{y \in \psi^m(x)} \psi(y)$. Notice that the function $f : gr\,\psi \to \Re$ is ψ-bounded from above w.r.t. the second argument, if $f(x, \cdot)$ is bounded from above on $\psi(x)$ for any $x \in X$.

Definition 3.15 We say that a p.s.d.s. $\varphi : X \to 2^X$ admits a generalized Lyapunov function if there exists a φ_*-bounded above w.r.t. to the second argument function $l : gr\,\varphi_* \to \Re_+$ such that

(L1) $l(x,y) \geq d(x,y)$, for any $x \in X,\ y \in \varphi(x)$

(L2) $l(x,z) \geq l(x,y) + l(y,z)$, for any $x \in X,\ y \in \varphi_*(x),\ z \in \varphi_*(y)$.

Any function satisfying the conditions indicated above (including φ_*-boundedness from above w.r.t. the second argument) is said to be a generalized Lyapunov function of p.s.d.s. φ.

Let $\bar{\Re} = \Re \cup \{+\infty\}$ be the extended real line. The least function possessing the properties (L1) and (L2) is the mapping $l_\varphi : gr\,\varphi_* \to \bar{\Re}$ constructed as follows. Set

$$T_\varphi^m(x,y) := \{\{x_r\}_{r=0}^m \mid x_0 = x, x_m = y, x_r \in \varphi(x_{r-1}), r = 1, \ldots, m\}$$

and denote by $T_\varphi^*(x,y)$ the collection of all finite trajectories of the p.s.d.s. φ connecting x and y :

$$T_\varphi^*(x,y) := \bigcup_{m=1}^{\infty} T_\varphi^m(x,y).$$

For $t = \{x_r\}_{r=0}^m \in T_\varphi^*(x,y)$ let $d(t) := \sum_{r=0}^{m-1} d(x_r, x_{r+1})$. Then l_φ is given by

$$l_\varphi(x,y) = \sup\{d(t) \mid t \in T_\varphi^*(x,y)\}, \quad (x,y) \in gr\,\varphi_*.$$

It is clear that φ admits a generalized Lyapunov function iff l_φ is φ_*-bounded above w.r.t. the second argument.

The existence of a generalized Lyapunov function does not yet guarantee the convergence of all trajectories of φ to its *final set*

$$E_\varphi := \{x \mid \varphi(x) = \{x\}\}.$$

At the same time, there exists, for lower semicontinuous p.s.d.s., a standard method of the formation of selectors ψ (as usual, we call a p.s.d.s. ψ *a selector of* φ if $\psi(x) \subseteq \varphi(x)$ for any $x \in X$) possessing the indicated property and having the same final set as φ. To give a slightly generalized version of the corresponding result from Maschler and Peleg (1976), note first that for any p.s.d.s. φ that admits a generalized Lyapunov function l, it holds (due to the φ_*-boundedness from above of l w.r.t. the second argument) that $\rho_\varphi(x) < \infty$ for any $x \in X$, where

$$\rho_\varphi(x) := \sup\{d(x,y) \mid y \in \varphi(x)\},\ x \in X.$$

Notice also that below in Proposition 3.16, there is no need to presuppose nonemptiness of the final set E_φ. On the contrary, nonemptiness of E_φ is one of the conclusions of this proposition.

Proposition 3.16 *Let φ be a lower semicontinuous p.s.d.s. on X admitting a generalized Lyapunov function. Then $E_\varphi \neq \emptyset$ and, moreover, for any $\delta \in (0,1)$ all trajectories of the p.s.d.s.*

$$\varphi_\delta(x) = \begin{cases} \{x\}, & x \in E_\varphi, \\ \{y \in \varphi(x) \mid d(x,y) > \delta\rho_\varphi(x)\}, & x \notin E_\varphi, \end{cases}$$

converge to elements of the final set E_φ.

Proof. Fix some $\delta \in (0,1)$. Since the p.s.d.s. φ_δ is correctly defined independently on whether the final set E_φ is empty or not, to prove the proposition it is sufficient to establish its second part, concerning the convergency of any trajectory of the p.s.d.s. φ_δ to the elements $x_* \in X$ satisfying equality $\varphi(x_*) = \{x_*\}$.

To do so, let us note, that from the definition of generalized Lyapunov function l it follows that for any trajectory $\{x_r\}_{r=0}^\infty$ of p.s.d.s. φ_δ and for any $m \geq 1$ we have: $\sum_{r=0}^m d(x_r, x_{r+1}) \leq l(x_0, x_{m+1})$. Hence, $\sum_{r=0}^\infty d(x_r, x_{r+1}) \leq \sup\{l(x_0, y) \mid y \in \varphi_*(x_0)\}$, which means, due to the completeness of X and φ_*-boundedness from above of l w.r.t. the second argument, that $\{x_r\}_{r=0}^\infty$ is convergent.

To prove that $x_* = \lim x_r \in E_\varphi$, notice that from the lower semicontinuity of φ it follows that the function $\rho = \rho_\varphi$ is lower semicontinuous: $M_\rho(a) = \{x \in X \mid \rho(x) \leq a\}$ is closed for any $a \in \Re$. Indeed, let $\{z_r\}_{r=1}^\infty$ be an arbitrary convergent sequence from $M_\rho(a)$ with $z = \lim z_r$. Fix some $y \in \varphi(z)$. By lower semicontinuity of φ there exists a convergent sequence $\{y_r\}_{r=1}^\infty$ such that $y = \lim y_r$ and $y_r \in \varphi(z_r)$ for any $r = 1, \ldots$. Since $d(y_r, z_r) \leq a$ for any $r = 1, \ldots$, we get $d(y,z) = \lim d(y_r, z_r) \leq a$ and therefore $\rho(z) = \sup\{d(y,z) \mid y \in \varphi(z)\} \leq a$.

To complete the proof, suppose on the contrary, that $x_* = \lim x_r$ does not belong to E_φ. As a result, we have $\rho(x_*) = c > 0$ and, consequently, by lower semicontinuity of ρ there exists some r_0 such that $\rho(x_r) > c/2$ for any $r \geq r_0$. By definition of φ_δ it follows that $d(x_r, x_{r+1}) > c\delta/2$ for any $r \geq r_0$, which contradicts to the convergency of the trajectory $\{x_r\}_{r=0}^\infty$. Hence $x_* = \lim x_r \in E_\varphi$ (and, as a by-product, $E_\varphi \neq \emptyset$). □

To prove the weak* external stability of the core $C(\alpha)$ from Proposition 3.16, it is enough to construct a lower semicontinuous selector ψ of the correspondence

$$\varphi^\alpha(x) := \alpha(x) \cup \{x\}, \quad x \in X,$$

which admits a generalized Lyapunov function and satisfies the relation $E_\psi = C(\alpha)$.

The following assertion is a suggestive argument for the construction of such systems $\psi \subseteq \varphi^\alpha$ (below, as usual, $\psi^{-1}(y) := \{x \in X \mid y \in \psi(x)\}$).

Proposition 3.17 *If p.s.d.s. φ satisfies the condition*

(C) For any $x \in X$ and for any $y \in \varphi(x) \setminus \{x\}$ it holds $x \in int\ \varphi^{-1}(y)$,

then for each $y \in \varphi_*(X)$ the function $l_\varphi(\cdot, y)$ is lower semicontinuous on the set $(\varphi_*)^{-1}(y) \setminus \{y\}$.

Proof. Let $y \in \varphi_*(X)$ and $l_\varphi(x, y) > a$ for some $x \in (\varphi_*)^{-1}(y) \setminus \{y\}$ and $a \in \Re$. By definition of l_φ there exists a finite sequence $x_0, x_1, \ldots, x_m$ such that $x_0 = x$, $x_m = y$, $x_{r+1} \in \varphi(x_r)$ for any $r = 0, 1, \ldots, m-1$ and, besides, $\sum_{r=0}^{m-1} d(x_r, x_{r+1}) > a$. W.l.g. we may assume that $x_1 \neq x_0$. Applying condition (C) we have: there exists a neighborhood U_x of x satisfying inclusion $U_x \subseteq \varphi^{-1}(x_1)$. Hence, from continuity of the function d it follows that there exists a neighborhood V_x of x such that $V_x \subseteq \varphi^{-1}(x_1) \subseteq (\varphi_*)^{-1}(y)$. Moreover we obtain

$$l_\varphi(z, y) \geq d(z, x_1) + \sum_{r=1}^{m-1} d(x_r, x_{r+1}) > a$$

for any $z \in V_x$. □

Remark 3.18 It is clear, that p.s.d.s. φ^α satisfies the condition (C) whenever $\alpha^{-1}(x)$ are open for any $x \in X$.

Let $\mathcal{K}$ be the collection of all subsets $K \subseteq X \times X$ satisfying the following conditions:

$(K1)$ For any $(x, y), (y, z) \in K$ it holds: $(x, z) \in K$;
$(K2)$ For any $(x, y) \in K$ with $x \neq y$ it holds: $\exists U_x[U_x \times \{y\} \subseteq K]$

(here and below U_x is a neighborhood of x).

We denote by $L(K)$ the family of bounded functions $l : K \to \Re_+$ such that $l(x, z) \geq l(x, y) + l(y, z)$, $(x, y), (y, z) \in K$, and, moreover, for any $y \in K(X)$ the function $l(\cdot, y)$ is lower semicontinuous on the set $K^{-1}(y) \setminus \{y\}$, where $K(X) = \{y \in X \mid \exists x \in X$ such that $(x, y) \in K\}$, and $K^{-1}(y) = \{x \in X \mid (x, y) \in K\}$. Let $\mathcal{L} := \bigcup_{K \in \mathcal{K}} L(K)$.

Considering as the desired selectors those systems that are of the form

$$\psi_l(x) := \{y \in \alpha(x) \mid l(x, y) > d(x, y)\} \cup \{x\}, \quad x \in X,$$

with $l \in \mathcal{L}$, we obtain the following weak* external stability test.

Theorem 3.19 *If the lower sections* $\alpha^{-1}(x)$ *are open for all* $x \in X$ *and, moreover, there exists a function* $l \in \mathcal{L}$ *such that*

$$\forall x \notin C(\alpha) \exists y \in \alpha(x) \text{ such that } l(x,y) > d(x,y),$$

then the core $C(\alpha)$ *is nonempty and weak* externally stable.*

Proof. Under the condition of the theorem there are a set $K \in \mathcal{K}$ and a function $l \in L(K)$ such that for any $x \notin C(\alpha)$ there exists y belonging to $\alpha(x) \cap K_l(x)$, where

$$K_l(x) := \{y \in X \mid (x,y) \in K \text{ and } l(x,y) > d(x,y)\}.$$

Put

$$\psi_l(x) := (\alpha(x) \cap K_l(x)) \cup \{x\}, \quad x \in X.$$

It is clear, that $\psi_l \subseteq \varphi^\alpha$ and, besides, $E_{\psi_l} = C(\alpha)$. Extending (if necessary) function l by the formula $l(x,x) := 0$, $x \in X$, we get: p.s.d.s. ψ_l admits a generalized Lyapunov function. According to Proposition 3.16, to complete the proof it remains to establish the lower semicontinuity of ψ_l. Let $x \in X, y \in \psi_l(x)$, and $x = \lim x_r$. If $y = x$, then $y = \lim y_r$ with $y_r = x_r$ for any $r = 1, \ldots$. Otherwise, we proceed as follows. Since $\alpha^{-1}(x)$ is open, and $l(\cdot, y)$ is lower semicontinuous on $K^{-1}(y) \setminus \{y\}$, there exists a neighborhood U_x of x such that $z \in \alpha^{-1}(y)$ and $l(z,y) - d(z,y) > 0$ for any $z \in U_x$. Hence, there exists $r_0 \geq 1$ such that $l(x_r, y) > d(x_r, y)$ and $y \in \alpha(x_r)$ for any $r > r_0$. Putting $y_r = x_r$ for $r \leq r_0$ and $y_r = y$ for $r > r_0$, we get: $y_r \in \psi_l(x_r)$ for any $r = 1, \ldots$, and $y = \lim y_r$. Thus, we have established that p.s.d.s. ψ_l is lower semicontinuous. □

It is easy to check that Theorem 3.19 remains valid if one only requires lower semicontinuity for φ^α, provided that $\alpha \subseteq K$ for some $K \in \mathcal{K}$ such that there exists $l \in L(K)$ which is jointly lower semicontinuous in its arguments. One of the most useful consequences of this fact is the following result.

Corollary 3.20 *Let* $\Gamma = (X, \alpha)$ *be an a.g. If* X *is a compact set and* φ^α *is lower semicontinuous, and, moreover, there exists a continuous function* $u : X \to \Re$ *satisfying*

$$\forall x \notin C(\alpha) \ \exists y \in \alpha(x) \text{ such that } u(x) - u(y) > d(x,y), \tag{3.5}$$

then the core of the game Γ *is nonempty and weak* externally stable.*

3.4 Accessibility of the core and weak* NM-solution

We consider now a simple specialization of the approach elaborated above to demonstrate accessibility of the cores in TU-case. Recall that, according to the terminology of Subsection 3.3 (see, also, Vasil'ev, 1984; Vasil'ev, 1987; and Makarov and Vasli'ev, 1989), the core $C(\alpha_v)$ of a TU-game v is accessible, if for any $x \in I(v)\backslash C(\alpha_v)$ there exists an α_v-monotone convergent sequence $\{x_r\}_{r=0}^{\infty}$ such that $x_0 = x$ and the limit imputation $x_* = \lim x_r$ belongs to the core $C(\alpha_v)$. To motivate our general setting, it should be mentioned first that the classic domination relation α_v, defined as above by

$$x \; \alpha_v \; y \iff \exists S \; [(y(S) \leq v(S)) \text{ and } (x_i < y_i, \; i \in S)],$$

is, generically, neither transitive, nor acyclic. Therefore, the dynamic system $x \mapsto \alpha_v(x) \cup \{x\}$, $x \in I(v)$, by no means directly leads to the core $C(\alpha_v)$. Quite to the contrary, even rather elementary examples of 3-person TU-cooperative games exhibit the fact, that imputation sets of these games contain open subsets such that no finite α_v-monotone path, originating from these subsets, reaches $C(\alpha_v)$.

Hence, except for some fairly exceptional situations (see, e.g. Bondareva, 1962, Kulakovskaya, 1971), in general setting it seems to make sense to speak at most about the limiting generalized NM-solution. Nevertheless, the approach given in the previous subsection allows us not only to establish the accessibility of the core of any balanced c.g. $v \in V$, but to construct, as well, the appropriate selectors of the p.s.d.s.

$$\varphi_v(x) = \alpha_v(x) \cup \{x\}, \quad x \in I(v),$$

whose final sets are globally stable and coincides with the core $C(\alpha_v)$. To show that the TU-property is essential for the accessibility of the core, an example of an NTU-cooperative game with a "black hole"[2] will also be given.

So, let v be an arbitrary c.g. v such that $C(\alpha_v) \neq \emptyset$. Since $I(v)$ is a compact set and lower sections $\alpha_v^-(x)$ are open for any $x \in I(v)$, to prove the accessibility of the core $C(\alpha_v)$ it is sufficient (due to Corollary 3.20) to construct a continuous function $u : I(v) \to \Re$, satisfying for some metric d on $I(v)$ the condition (3.20) for $\alpha = \alpha_v$. To this end fix the Euclidean distance $d(x,y) := ||x-y||_2 = \left(\sum_N (x_i - y_i)^2\right)^{1/2}$ as a metric on $I(v)$ and put

$$u_v(x) = 4nd(x, C(\alpha_v)), \quad x \in I(v),$$

[2] As usual, the latter is regarded to be a closed subset of imputations that doesn't intersect the core and contains all the monotone trajectories originating from it.

where, as usual, $d(x, Y) := \inf\{d(x,y) \mid y \in Y\}$ denotes the distance (in metric d) from x to Y.
Note that, due to the compactness of $C(\alpha_v)$, the function u_v is continuous. In addition, the following key assertion holds.

Proposition 3.21 *For any $x \in I(v) \setminus C(\alpha_v)$ there exists an imputation $z \in \alpha_v(x)$ such that*

$$u_v(x) - u_v(z) > ||x - z||_2. \tag{3.6}$$

Proof. Let us fix an arbitrary imputation $x \in I(v) \setminus C(\alpha_v)$ and show that there exists $z \in I(v)$ satisfying inequality (3.6) and the following weak domination condition: $z \in \beta_v(x)$, where

$$x \ \beta_v \ z \iff \exists S \ [(x(S) < z(S) \leq v(S)) \text{ and } (x_i \leq z_i, \ i \in S)].$$

It is clear that $x \ \beta_v \ z$ implies that every neighborhood of z contains an element $y \in I(v)$ which dominates x w.r.t. α_v. Observe also, that due to the above-mentioned continuity of u_v, we have that the inequality (3.6) holds in some neighborhood of z. Therefore the existence of z, satisfying relation $x \ \beta_v \ z$ and inequality (3.6), proves our proposition.

We may assume w.l.g. that $v(i) = 0$ for every $i \in N$. Consider the function $\bar{v}(S) := \min\{v(S), v(N)\}$, $S \subseteq N$. Since, clearly, $\alpha_{\bar{v}} \subseteq \alpha_v$, $I(\bar{v}) = I(v)$, and $C(\alpha_v) = C(\alpha_{\bar{v}}) = C(\bar{v})$, it is sufficient to restrict ourselves to the case that $v(S) \leq v(N)$ for any $S \subseteq N$. Hence, we may assume without loss of generality that

$$C(\alpha_v) = \{x \in I(v) \mid x(S) \geq v(S), \quad S \subseteq N\}. \tag{3.7}$$

In order to construct z, consider the imputation $y \in C(v)$, closest to x w.r.t. the norm $|| \cdot ||_2$. It is not very hard to verify that there exists a nonempty coalition $S \subseteq N$ such that $x(S) < y(S) = v(S)$. To do so, put $\Omega_x = \{T \subseteq N \mid x(T) < v(T)\}$. To prove that there exists a coalition $S \in \Omega_x$ such that $y(S) = v(S)$, suppose on the contrary (since $y \in C(v)$) that $y(T) > v(T)$ for any $T \in \Omega_x$. Then these inequalities imply that $y_\lambda = \lambda x + (1 - \lambda)y$ belongs to $C(v)$ with $\lambda \in (0, 1)$ small enough. Since, obviously, $||x - y_\lambda||_2 < ||x - y||_2$, this gives a contradiction.

Now, fix the above-mentioned coalition S and put

$$S_1 = \{i \in S | \ x_i < y_i\}, \ S_2 = \{i \in N \setminus S | \ x_i > y_i\},$$

$$T_1 = \{i \in S | \ x_i \geq y_i\}, \ T_2 = \{i \in N \setminus S \mid x_i \leq y_i\}.$$

By definition of S, x, and y, the coalitions S_1 and S_2 are not empty. Therefore, if $T_1^+ = \{i \in T_1 | \ x_i > y_i\} = \emptyset$, then y is an imputation satisfying the properties stated in the proposition. In case $T_1^+ \neq \emptyset$, then setting

$$e(x,S) = v(S) - x(S),$$
$$q_j = e(x,S)/|y(S_j) - x(S_j)|, \ j = 1,2,$$

we define $z \in \Re^N$ by the formula

$$z_i = \begin{cases} x_i + q_1(y_i - x_i), & \text{if } i \in S_1, \\ x_i + q_2(y_i - x_i), & \text{if } i \in S_2, \\ x_i, & \text{if } i \in T_1 \cup T_2. \end{cases}$$

We now have that $z \in I(v)$. To prove this, notice that

$$e(x,S) = |\sum_{i \in S_j \cup T_j} (y_i - x_i)|, \ j = 1,2, \tag{3.8}$$

implies $q_1, q_2 \in (0,1]$, and, hence, $z_i \geq \min\{x_i, y_i\} \geq 0$ for any $i \in S_1 \cup S_2$. This, together with the equalities $e(x,S) = q_j|y(S_j) - x(S_j)|$, $j = 1,2$, proves that $z \in I(v)$. Taking into account that $S_1 \neq \emptyset$ and $z(S) = x(S) + e(x,S) = v(S)$, $z_i = x_i$, $i \in T_1$, we have that $x(S) < z(S) \leq v(S)$, $x_i \leq z_i$, each $i \in S$. By definition of β_v, these inequalities imply $x \ \beta_v \ z$ with respect to the coalition $S = S_1 \cup T_1$.

In order to check the inequality (3.6), note that by definition of z and q_1, q_2, we get an upper bound for $||x - z||_2$:

$$||x - z||_2 = \left[\sum_{j=1}^{2} q_j^2 (\sum_{i \in S_j} (y_i - x_i)^2)\right]^{1/2} =$$

$$= e(x,S) \left[\sum_{j=1}^{2} \sum_{i \in S_j} (y_i - x_i)^2 / (y(S_j) - x(S_j))^2\right]^{1/2} \leq \sqrt{2}\, e(x,S).$$

Further, by definition of u_v we have

$$(4n)^{-1}(u_v(x) - u_v(z)) \geq ||x - y||_2 - ||z - y||_2. \tag{3.9}$$

To get a lower bound for the difference $||x - y||_2 - ||z - y||_2$, observe that by definition of z the equality $||z - y||_2 = [(||x - y||_2)^2 - (Q_1 + Q_2)]^{1/2}$ is valid, where

$$Q_j = (2q_j - q_j^2) \cdot \sum_{i \in S_j} (x_i - y_i)^2, \ j = 1,2.$$

Since $q_1, q_2 \in (0,1]$, we get $2q_1 - q_1^2 > 0,\ 2q_2 - q_2^2 > 0$ and, consequently, $Q_1 > 0, Q_2 > 0$. Therefore

$$\begin{aligned}||x-y||_2 - ||z-y||_2 &= \frac{(||x-y||_2)^2 - (||z-y||_2)^2}{||x-y||_2 + ||z-y||_2} \\ &\geq (Q_1+Q_2)/2||x-y||_2. \end{aligned} \tag{3.10}$$

To estimate the right hand side of (3.10) we apply the special case of the well-known Cauchy-Schwartz-Bunyakovskii inequality

$$\sum_{i=1}^{m} a_i^2/(\sum_{i=1}^{m} a_i)^2 \geq 1/m. \tag{3.11}$$

To be more detailed, since the equalities (3.8) and the definition of T_1, T_2 yield

$$\sum_{i \in T_j} (x_i - y_i)^2 \leq [x(T_j) - y(T_j)]^2 \leq [y(S_j) - x(S_j)]^2, \quad j = 1, 2,$$

by (3.11) we obtain

$$\begin{aligned}\sum_{i \in T_j} (x_i - y_i)^2 &\leq [y(S_j) - x(S_j)]^2 \\ &\leq |S_j| \sum_{i \in S_j} (x_i - y_i)^2, \quad j = 1, 2. \end{aligned} \tag{3.12}$$

Hence, due to the inequalities $|S_j| < n,\ j = 1, 2$, we obtain

$$\sum_{i \in S_j \cup T_j} (x_i - y_i)^2 \leq n \sum_{i \in S_j} (x_i - y_i)^2, \quad j = 1, 2.$$

Summing up and extracting the square roots, we get the inequality

$$||x-y||_2 \leq \sqrt{n} \left[\sum_{i \in S_1 \cup S_2} (x_i - y_i)^2 \right]^{1/2},$$

which yields, by definition of Q_1, Q_2, that

$$||x-y||_2 \leq \sqrt{n} \left[\sum_{j=1}^{2} Q_j/(2q_j - q_j^2) \right]^{1/2}.$$

Combining the latter inequality and estimation (3.10) we have

$$||x-y||_2 - ||z-y||_2 \geq \frac{Q_1+Q_2}{2\sqrt{n}\left[\sum_{j=1}^{2} Q_j/(2q_j - q_j^2)\right]^{1/2}}. \tag{3.13}$$

To evaluate the denominator in the right hand side of (3.13) we rewrite (using the definition of q_1, q_2) the last inequalities in (3.12) as

$$q_j^2 \sum_{i\in S_j} (x_i - y_i)^2 \geq e^2(x,S)/|S_j|, \quad j = 1,2.$$

From here, taking into account the definition of Q_1, Q_2 and inequalities $2q_j - q_j^2 \geq q_j$, $j = 1,2$, we obtain

$$2q_j - q_j^2 \geq e^2(x,S)/(|S_j|Q_j), \quad j = 1,2.$$

Therefore, we may proceed with evaluation of the above mentioned denominator

$$2\sqrt{n}\left[\sum_{j=1}^{2} Q_j/(2q_j - q_j^2)\right]^{1/2} \leq 2\sqrt{n}\left(\sum_{j=1}^{2} |S_j|Q_j^2\right)^{1/2} /e(x,S) \leq$$

$$\leq 2n(Q_1^2 + Q_2^2)^{1/2}/e(x,S) \leq 2n(Q_1+Q_2)/e(x,S).$$

Finally, applying (3.9) and (3.13) we get $u_v(x) - u_v(z) \geq 2e(x,S)$ and, using the upper bound $||x-z||_2 \leq \sqrt{2}\, e(x,S)$ as established above, we obtain the desired inequality $u_v(x) - u_v(z) > ||x-z||_2$. □

Taking into account continuity of the functions u_v, $v \in V$, Corollary 3.20 and Proposition 3.21, we obtain one of the main results of the paper.

Theorem 3.22 *The core $C(\alpha_v)$ of any TU-cooperative game v is accessible whenever it is nonempty.*

Let us recall once more (von Neumann and Morgenstern, 1953) that $Y \subseteq X$ is internally stable w.r.t. α, if $(x,y) \notin \alpha$ for every $x,y \in Y$ such that $x \neq y$. An internally and externally stable (w.r.t. α) subset $Y \subseteq X$ is called an NM-solution of the a.g. $\Gamma = (X, \alpha)$.

Definition 3.23 A subset $Y \subseteq I(v)$ is a weak* NM-solution of a c.g. v if it is an internally and externally stable w.r.t. the limit transitive closure α_v^* of the classic domination relation α_v.

Theorem 3.24 *For any TU-cooperative game v there exists a weak* NM-solution. This solution is unique and coincides with the core $C(\alpha_v)$, if the latter is not empty, and may be chosen finite, if $C(\alpha_v) = \emptyset$.*

Proof. From Theorem 3.22 which states that $C(\alpha_v)$ is externally stable w.r.t. α_v^* (in case $C(\alpha_v) \neq \emptyset$) it follows that the core $C(\alpha_v)$ is the unique weak* NM-solution of c.g. v, when nonempty. To analyze what happens when $C(\alpha_v) = \emptyset$, we apply the argumentation similar to that in the proof of Theorem 3.7. Put $\mathcal{A}(y) := (\alpha_v^*)^{-1}(y),\ y \in I(v)$. Since the core of v coincides with the set of α_v^*-maximal elements on $I(v)$, it follows that in case $C(\alpha_v) = \emptyset$ we have that $I(v) = \bigcup_{y \in I(v)} \mathcal{A}(y)$. As it was already mentioned, from openness of $\alpha_v^{-1}(x)$ for any $x \in I(v)$ it follows that $\mathcal{A}(y)$ are open for all $y \in I(v)$. Taking into account that $I(v)$ is a compact set, we may choose some finite set $Y = \{y_1, \ldots, y_m\}$ such that

$$I(v) = \bigcup_{i=1}^{m} \mathcal{A}(y_i). \tag{3.14}$$

From Lemma 3.8 and the transitivity of α_v^* it follows that there exists (an ordinary) NM-solution Y_0 of the finite abstract game $\Gamma = (Y, \beta)$, where $\beta := \alpha_v^* \bigcap Y \times Y$. Now, taking into account equality (3.14), it is not very hard to verify that the finite set Y_0 is a weak* NM-solution of TU-cooperative game v. □

To conclude this section, we show that the TU-property is essential for the core to be a weak* NM-solution by giving an example of an NTU-cooperative game with a "black hole"[3]. More precisely, a 3-person NTU-cooperative game G is constructed with $G(N)$ containing a closed subset B such that $B \cap C(\alpha_G) = \emptyset$ and all monotone trajectories originating from any $x \in B$ stay in B. It is worth to note that such a phenomenon (presence of a " black hole ") is stable w.r.t. any small enough variation of the game under consideration.

Take $N = \{1, 2, 3\}$ and let the characteristic function G be given by

$$\begin{aligned} G(\{1,2,3\}) &= \{x \in \Re_+^{\{1,2,3\}} | \sum_{i=1}^{3} x_i = 1\}, \\ G(\{1,2\}) &= \{x \in \Re_+^{\{1,2\}} | x_1 + x_2 \leq 2/3\}, \\ G(\{1,3\}) &= \{x \in \Re_+^{\{1,3\}} | x_1 + x_3 \leq 2/3,\ x_1 \leq 1/3\}, \end{aligned}$$

[3]I owe to Dr. Batanin M.A. for helpful discussion on the subject.

$$\begin{aligned} G(\{2,3\}) &= \{x \in \Re_+^{\{2,3\}} | x_2 + x_3 \leq 5/6,\ x_2 \leq 1/2\}, \\ G(\{i\}) &= \{0\},\ i \in N. \end{aligned}$$

Further, we apply the standard domination relation α_G on $G(N)$, defined by

$$x\ \alpha_G\ y \Leftrightarrow \exists S \subseteq N[\forall i \in S(x_i < y_i) \text{ and } (y_S \in G(S))],$$

where y_S is the restriction of y to S. It is not hard to verify that the core $C(\alpha_G)$, which consists of the set of α_G-maximal elements of $G(N)$, is given by

$$C(\alpha_G) = \{x \in G(N) | x_1 \geq 1/3,\ x_2 \geq 1/2\}.$$

Defining

$$B = \{x \in G(N) | x_1 + x_2 \leq 2/3,\ x_1 + x_3 \leq 2/3,\ x_2 + x_3 \leq 5/6\},$$

we prove that B meets the requirements: (i) B is closed, (ii) $B \cap C(\alpha_G) = \emptyset$ and (iii) for any $y \in G(N)$ with $x\ \alpha_G^*\ y$ for some $x \in B$ it holds that y belongs to B.

First, closedness of B is evident. Second, since the inequalities

$$x_1 + x_2 \leq 2/3,\ x_1 \geq 1/3,\ x_2 \geq 1/2$$

are inconsistent, it is clear that $B \cap C(\alpha_G) = \emptyset$. Finally, as B is closed, the only thing we need to prove that B satisfies (iii) is the implication

$$(x\ \alpha_G\ y) \text{ and } (x \in B) \Rightarrow y \in B.$$

Fix an arbitrary $x \in B$ and some $y \in G(N)$ such that $x\ \alpha_G\ y$. To prove $y \in B$ let us look through three possible situations which correspond to the two-person blocking coalitions $S = \{i, j\}$ (such that $(y_i, y_j) \in G(\{i, j\})$ and $x_i < y_i$, $x_j < y_j$).

1. $S = \{1, 2\}$. Since in this case $y_1 + y_2 \leq 2/3$, it remains to prove that

$$y_1 + y_3 \leq 2/3,\ y_2 + y_3 \leq 5/6.$$

Assuming $y_1 + y_3 > 2/3$, we obtain $y_2 < 1/3$, which contradicts $y_2 > x_2 \geq 1/3$. Assuming $y_2 + y_3 > 5/6$ we obtain $y_1 < 1/6$, contradicting $y_1 > x_1 \geq 1/6$ (since $x_2 + x_3 \leq 5/6$).

2. $S = \{1, 3\}$. By definition of $G(\{1, 3\})$ it holds

$$y_1 + y_3 \leq 2/3,\ y_1 \leq 1/3.$$

Since $x \in B$ implies $x_1 \geq 1/6$, $x_3 \geq 1/3$, from $y_1 > x_1 \geq 1/6$ we get $y_2 + y_3 < 5/6$. In the same manner, from $y_3 > x_3 \geq 1/3$ we have $y_1 + y_2 \leq 2/3$.

3. $S = \{2, 3\}$. By definition of $G(\{2, 3\})$ it follows

$$y_2 + y_3 \leq 5/6,\ y_2 \leq 1/2.$$

Since $x \in B$ implies $x_2 \geq 1/3$, $x_3 \geq 1/3$, from $y_2 > x_2 \geq 1/3$ we obtain $y_1+y_3 < 2/3$. By applying inequalities $y_3 > x_3 \geq 1/3$ we have: $y_1+y_2 < 2/3$.

So, in all possible situations $x\ \alpha_G\ y$ and $x \in B$ imply $y \in B$. Hence, bearing in mind that B is closed, we have that from $x\ \alpha_G^*\ y$ and $x \in B$ it always follows that y belongs to B. To complete the proof that the core $C(\alpha_G)$ *is not outer stable w.r.t.* α_G^*, notice that $B \neq \emptyset$ and $B \cap C(\alpha_G) = \emptyset$.

One can easily verify that any small enough perturbation of this game of bounded side payments preserves unstability of the core $C(\alpha_G)$.

4 Totally contractual allocations and equilibrium

Below, an equilibrium characterization of the so-called totally contractual cores (see Makarov, 1982; Vasil'ev, 1984) in some classes of competitive economies is given. Because of quite complicated logical structure of the contractual domination relations we pay strong attention to the pure descriptive aspects of the concepts under consideration. Rather simple sufficient conditions, guaranteeing coincidence of the totally contractual core and the set of Walrasian equilibrium allocations, are established, and the structure of domination relations, induced by several rules of breaking the contracts, is studied. The game-theoretic approach in this part of the paper is based on the reduction of the contractual blocking rules to more simple domination relations in cooperative games, associated with the original blocking in question.

4.1 Totally contractual set and M-core

We consider a slightly generalized pure exchange model

$$\mathcal{E} = \langle N, \{X_i, w^i, \alpha_i\}_{i \in N}, \Omega \rangle, \tag{4.1}$$

where $N := \{1, \ldots, n\}$ is a set of consumers, and $X_i \subseteq \Re^l$, $w^i \in \Re^l$, $\alpha_i \subseteq X_i \times X_i$ are their consumption sets, initial endowments, and individual preference relations, respectively. Further Ω, is a nonempty subset of 2^N, called a coalitional structure of $\mathcal{E}$, yielding a collection of nonempty admissible coalitions $S \subseteq N$, which may join efforts of their participants in order to improve (block) any current distribution of the total initial endowment $\sum_{i \in N} w^i$.

To present the main definition of contractual domination (blocking) in $\mathcal{E}$, we modify first the relevant definitions from Makarov (1982), aiming to clarify some features of the elementary interchange structure we deal with. For each $S \in \Omega$ we fix some subset $M_S \subseteq \Re^l$ such that $0 \in M_S$, and put $M_\Omega := \{M_S\}_{S \in \Omega}$.

Definition 4.1 A contract (of type M_S) of coalition $S \in \Omega$ is a collection of vectors $\mathrm{v} = \{\mathrm{v}^{ij}\}_{i,j \in S}$ satisfying (i) $\mathrm{v}^{ii} = 0$, for each $i \in S$, (ii) $\mathrm{v}^{ij} \in M_S$, for each $i, j \in S$, and (iii) $\mathrm{v}^{ij} = -\mathrm{v}^{ji}$, for each $i, j \in S$.

Coalition S entering into a contract v will be denoted by $S(\mathrm{v})$, as well.

Components of a vector v^{ij}, appearing in definition of the contract $\mathrm{v} = \{\mathrm{v}^{ij}\}_{i,j \in S}$, indicate amount of the corresponding commodities used in the bilateral exchange between the participants $i, j \in S(\mathrm{v})$. A nonnegative component $\mathrm{v}_s^{ij} \geq 0$ of v^{ij} denotes the amount of commodity s that agent j is obliged to deliver to agent i and the absolute value of a negative component $\mathrm{v}_t^{ij} < 0$ measures the amount of commodity t agent i has to deliver to agent j.

The subsets M_S define admissible types of elementary exchanges within the contract $\mathrm{v} = \{\mathrm{v}^{ij}\}_{i,j \in S}$, e.g., in case $M_S := \{x \in \Re^l \mid p^S \cdot x = 0\}$ the only constraint concerning v^{ij}, $i, j \in S$, is that the bilateral exchanges v^{ij} should be equivalent w.r.t. the (fixed) prices p^S for coalition S.

We call $\mathrm{v} = \{\mathrm{v}^{ij}\}_{i,j \in S}$ a proper contract, if either v is a trivial contract (i.e., $\mathrm{v}^{ij} = 0$ for all $i, j \in S(\mathrm{v})$), or for any $i \in S(\mathrm{v})$ there exist $j, k \in S(\mathrm{v})$ such that v^{ij} contains strictly positive components, and v^{ik} contains strictly negative components. Note, that the properness of a contract v makes it possible to break it by each member $i \in S(\mathrm{v})$, simply by not delivering the corresponding commodity to the participant $k = k(i)$. Any finite set $\mathcal{V} = \{\mathrm{v}_r\}_{r \in \mathcal{R}}$, consisting of proper contracts v_r, is called a (proper) contract system of type M_Ω (contractual M_Ω-system, or c.s. (of type M_Ω), for short). We want to stress that the elements of $\mathcal{R}$ are supposed to be "titles" of any type (not necessary natural numbers), naming the contracts belonging to the set $\mathcal{V}$.

Thus, it is supposed that the members of any coalition $S \in \Omega$ may enter several contracts. Moreover, a c.s. $\mathcal{V}$ may contain several samples of the same contract v differing just by their titles (or, by their names, to be exact). To simplify the terms, we call a c.s. $\widetilde{\mathcal{V}} = \{\widetilde{\mathrm{v}}_r\}_{r \in \widetilde{\mathcal{R}}}$ a subsystem of c.s. $\mathcal{V} = \{\mathrm{v}_r\}_{r\in\mathcal{R}}$ iff $\widetilde{\mathcal{R}} \subseteq \mathcal{R}$, and $\widetilde{\mathrm{v}}_r = \mathrm{v}_r$ for any $r \in \widetilde{\mathcal{R}}$. Hence, we consider contract systems $\mathcal{V}$ and $\widetilde{\mathcal{V}}$ to be identical iff $\mathcal{R} = \widetilde{\mathcal{R}}$ and $\widetilde{\mathrm{v}}_r = \mathrm{v}_r$ for any $r \in \mathcal{R}$.

Denote by $\Delta^i(\mathcal{V})$ the net outcome of agent i resulting after a c.s. $\mathcal{V} = \{\mathrm{v}_r\}_{r\in\mathcal{R}}$ is accepted and realized

$$\Delta^i(\mathcal{V}) := \begin{cases} \sum_{r|i\in S(\mathrm{v}_r)} \sum_{j\in S(\mathrm{v}_r)} \mathrm{v}_r^{ij}, & i \in \bigcup_{r\in\mathcal{R}} S(\mathrm{v}_r), \\ 0, & \text{otherwise.} \end{cases}$$

Since $\mathrm{v}_r^{ij} = -\mathrm{v}_r^{ji}$ for any $i, j \in S(\mathrm{v}_r)$, $r \in \mathcal{R}$, we have that $\sum_{i\in N} \Delta^i(\mathcal{V}) = 0$ and, hence, $\sum_{i\in N} x^i(\mathcal{V}) = \sum_{i\in N} w^i$ with

$$x^i(\mathcal{V}) = x^i(\mathcal{V}, \mathcal{E}) := w^i + \Delta^i(\mathcal{V}), \quad i \in N,$$

to be the resulting outcome of the agent i, obtained by entering c.s. $\mathcal{V}$. In the sequel we call $x(\mathcal{V}) = x(\mathcal{V}, \mathcal{E}) := (x^i(\mathcal{V}, \mathcal{E}))_{i\in N}$ a contractual M_Ω-allocation of the economy $\mathcal{E}$ (c.a. (of type M_Ω), for short).

Definition 4.2 A contractual system $\mathcal{V}$ is called feasible if $x^i(\mathcal{V}) \in X_i$ for any $i \in N$.

By definition, feasibility of a c.s. $\mathcal{V}$ guarantees a contractual M_Ω-allocation $x(\mathcal{V})$ to belong to the set

$$X(N) = X_\mathcal{E}(N) := \{(x^i)_{i\in N} \in \prod_{i\in N} X_i \mid \sum_{i\in N} x^i = \sum_{i\in N} w^i\}$$

of attainable allocations of the economy $\mathcal{E}$.

Recall, that due to the properness, any contract $\mathrm{v} \in \mathcal{V}$ may, in principal, be broken by any participant $i \in S(\mathrm{v})$. To present a formal description of the outcomes of rescinding (breaking) some contracts v_r, $r \in \mathcal{R}'$ with $\mathcal{R}' \subseteq \mathcal{R}$, we denote by $A(\mathcal{V}, \mathcal{R}', \mathcal{E})$ a collection of all feasible contract systems $\widetilde{\mathcal{V}} = \{\mathrm{v}_r\}_{r\in\widetilde{\mathcal{R}}}$ of the model $\mathcal{E}$, satisfying (i) $\widetilde{\mathcal{V}}$ is a subsystem of $\mathcal{V}$, (ii) $\widetilde{\mathcal{R}} \subseteq \mathcal{R} \setminus \mathcal{R}'$, (iii) $\widetilde{\mathcal{V}}$ is the maximal (by inclusion) subsystem of $\mathcal{V}$, satisfying condition (ii). Thus, $A(\mathcal{V}, \mathcal{R}', \mathcal{E})$ contains all the outcomes of the breaking (cancellation) procedure, consisting of two steps. At the first step we nullify all the contracts v_r, $r \in \mathcal{R}'$. In case $\mathcal{U} := \{\mathrm{v}_r\}_{r\in\mathcal{R}\setminus\mathcal{R}'}$ is feasible we put $A(\mathcal{V}, \mathcal{R}', \mathcal{E}) := \{\mathcal{U}\}$. Otherwise, at the second, final step, we continue to nullify the contracts with the titles belonging to some subset $\mathcal{R}''$ of $\mathcal{R} \setminus \mathcal{R}'$ in order to provide the feasibility of the subsystem $\tilde{\mathcal{U}} = \{\mathrm{v}_r\}_{r\in\mathcal{R}\setminus(\mathcal{R}'\cup\mathcal{R}'')}$. The only requirement concerning $\mathcal{R}''$ is its minimality: according to condition

(iii) we cancel an as "small" as possible collection of contracts (belonging to $\mathcal{U}$), provided that not cancelled ones constitute a feasible contractual system. Speaking differently, at the second step we choose some maximal (by inclusion) feasible subsystem of $\mathcal{V}$ with contractual titles belonging to $\mathcal{R}\backslash\mathcal{R}'$. Since there may be either none, or several nonempty feasible subsystems of $\mathcal{V}$ of that type, the collection $A(\mathcal{V}, \mathcal{R}', \mathcal{E})$ may be either trivial (i.e., consisting of the empty subsystem), or it may contain several nonempty feasible subsystems of $\mathcal{V}$.

Remark 4.3 Certainly, it might be reasonable to consider some other "rescinding rules", e.g., the rule, which allows to cancel the contracts from $\mathcal{U} = \{\mathrm{v}_r\}_{r\in\mathcal{R}\backslash\mathcal{R}'}$ step by step according to some ordering on $\mathcal{R} \setminus \mathcal{R}'$. This cancellation process should last until those contracts, which are not cancelled yet, constitute (for the first time) a feasible subsystem of $\mathcal{U}$. Other rules may allow either to take any feasible subsystem of $\mathcal{U}$, or to take $\mathcal{R}'$ from the very beginning in such a way that $\mathcal{U}$ is feasible (provided, probably, that $\mathcal{R}'$ satisfies some a priori given assumptions). We plan to elaborate on these and other approaches in further research.

To describe which way a coalition $S \in \Omega$ can improve upon a c.a. $x(\mathcal{V})$, generated by a feasible c.s. $\mathcal{V} = \{\mathrm{v}_r\}_{r\in\mathcal{R}}$ of type M_Ω, we suppose first that together with the possibility to cancel some contracts v_r with $r \in \mathcal{R}^S = \mathcal{R}^S_{\mathcal{V}} := \{r \in \mathcal{R} \mid S(\mathrm{v}_r)\bigcap S \neq \emptyset\}$, it is allowed for the coalition S to enter some new contract w. Denote by $\mathcal{E}_{\mathrm{w}}$ the modification of $\mathcal{E}$ generated by w:

$$\mathcal{E}_{\mathrm{w}} = \langle N, \{X_i, w^i + \Delta^i(\mathrm{w}), \alpha_i\}_N, \Omega\rangle,$$

where, as before, $\Delta^i(\mathrm{w}) := \Delta^i(\{\mathrm{w}\})$, i.e.,

$$\Delta^i(\mathrm{w}) = \begin{cases} \sum_{j\in S(\mathrm{w})} \mathrm{w}^{ij}, & i \in S(\mathrm{w}), \\ 0, & \text{otherwise.} \end{cases}$$

Definition 4.4 A coalition $S \in \Omega$ improves (blocks) a feasible contractual M_Ω-system $\mathcal{V} = \{\mathrm{v}_r\}_{r\in\mathcal{R}}$, if there exists a subset $\mathcal{R}' \subseteq \mathcal{R}^S$, a proper contract w of type M_S, and a system $\mathcal{V}' \in A(\mathcal{V}, \mathcal{R}', \mathcal{E}_{\mathrm{w}})$ such that $x^i(\mathcal{V}, \mathcal{E})\ \alpha_i\ x^i(\mathcal{V}', \mathcal{E}_{\mathrm{w}})$ for any $i \in S$, with $x^i(\mathcal{V}', \mathcal{E}_{\mathrm{w}}) \in \mathcal{P}_i(x^i(\mathcal{V}, \mathcal{E}))$ for at least one $i \in S$.

(A usual, $\mathcal{P}_i(z) := \{x^i \in \alpha_i(z) \mid (x^i, z) \notin \alpha_i\}$ denotes the set of those bundles from X_i, which are strictly preferred to z.)

Definition 4.5 A feasible contractual M_Ω-system $\mathcal{V}$ is called quasi-stable, if there is no coalition $S \in \Omega$, which improves $\mathcal{V}$.

For any $x \in X(N)$ denote by $\mathcal{W}(x) = \mathcal{W}_{M_\Omega}(x)$ the set of all c.s. $\mathcal{V}$ of type M_Ω such that $x = x(\mathcal{V})$. To isolate the allocations $x \in X(N)$ with $\mathcal{W}_{M_\Omega}(x) \neq \emptyset$, whose coalitional stability is independent on the concrete representation via some c.s. $\mathcal{V}$ of type M_Ω, we introduce the next definition.

Definition 4.6 An allocation $x \in X(N)$ is called a totally contractual M_Ω-allocation (shortly t.c.a. (of type M_Ω)), if $\mathcal{W}_{M_\Omega}(x)$ is a nonempty set, and it contains only quasi-stable contractual systems (i.e., for every $\mathcal{V} \in \mathcal{W}_{M_\Omega}(x)$ it holds: $\mathcal{V}$ is quasi-stable).

The set $D_0^{M_\Omega}(\mathcal{E})$ of all the totally contractual M_Ω-allocations of $\mathcal{E}$ is called the totally contractual set (of type M_Ω) of economy $\mathcal{E}$.

Note that any coalitional structure Ω admits the (unique) representation as the union of (locally) irreducible coalitional substructures, inscribed into the corresponding components of N. Therefore in what follows we suppose Ω to be irreducible itself (recall (Vasil'ev, 1984), that coalitional structure Ω is called irreducible, if for any nontrivial partition $\{N_1, N_2\}$ of N there is $S \in \Omega$ such that $S \bigcap N_i \neq \emptyset$, $i = 1, 2$; as usual, a partition $\{N_1, N_2\}$ is called nontrivial if $N_1, N_2 \neq \emptyset$). In the sequel it is assumed that all the subsets M_S are identical and equal to some linear subspace $M \subseteq \Re^l$ satisfying the following sign-assumption: every nonzero $z \in M$ contains both strictly positive and strictly negative components.

Remark 4.7 From the Bipolar Theorem and Minkovski Separation Theorem it follows immediately that a linear subspace $M \neq \Re^l$ satisfies the above mentioned sign-assumption iff its polar subspace $M^0 := \{p \in \Re^l \mid p \cdot z = 0,\ z \in M\}$ meets the requirement

$$M^0 \bigcap \text{int } \Re^l_+ \neq \emptyset. \tag{4.2}$$

It is clear that in this case there exists a nonempty finite subset $P_M \subseteq \Re^l$, containing at least one strictly positive vector, such that $M = \{z \in \Re^l \mid p \cdot z = 0,\ p \in P_M\}$. Hence, the economical meaning of the constraint imposed on the type of the contracts we consider below is as follows: elementary exchange bundles v^{ij} should be compatible with all the fixed-price vectors p, belonging to some nonempty finite subset $P_M \subseteq \Re^l$ that contains at least one strictly positive price vector $\bar{p}$.

Under the sign-assumption it also can easily be shown that for any $S \subseteq N$ with $|S| \geq 3$ there always exists a proper "zero-contract" w of type M (i.e., a proper contract $\mathrm{w} \neq 0$ of type M such that (i) $S(\mathrm{w}) = S$, and (ii) $\Delta^i(\mathrm{w}) = 0$ for any $i \in S$). It means that the properness of any contract v of type $M \neq$

$\{0\}$ with $|S(\mathrm{v})| \geq 3$ is guaranteed "automatically": $\mathrm{u} := \mathrm{v} + \lambda\mathrm{w}$ is a proper contract of type M with $S(\mathrm{u}) = \mathrm{S}(\mathrm{v})$ and $\Delta^i(\mathrm{u}) = \Delta^i(\mathrm{v})$, $i \in N$, provided that w is a proper "zero-contract" of type M with $\mathrm{w} \neq 0$, $S(\mathrm{w}) = \mathrm{S}(\mathrm{v})$, and $\lambda > 0$ is large enough. Hence, everywhere below we may (and will) deal with some contracts and contract systems of type M, not necessarily satisfying the properness-assumption for not-two-person coalitions (at least, in those situations, where only properness "modulo zero-contract" matters).

Denote by $\mathcal{W}_M$ the collection of all c.s. of type M_Ω with $M_S = M$ for any $S \in \Omega$. In what follows, the contracts v of type $M_S = M$, as well as the systems $\mathcal{V} \in \mathcal{W}_M$, and allocations $x = x(\mathcal{V})$, will be called M-contracts, M-systems and M-allocations, respectively. To present more detailed description of some undominated (in the sense of Definition 4.4) M-allocations $x(\mathcal{V})$, we characterize first all feasible allocations, attainable by entering into the contract systems of type M. Put

$$M_{\mathcal{E}}(N) := \{x \in \prod_{i \in N} X_i \mid \exists \mathcal{V} \in \mathcal{W}_M : x = x(\mathcal{V})\}.$$

Proposition 4.8 *If Ω is irreducible coalition structure, then*

$$M_{\mathcal{E}}(N) = \{x \in X(N) \mid \exists \Delta \in \Delta_M(N) : x = w + \Delta\},$$

where $\Delta_M(N) := \{\Delta = (\Delta^i)_{i\in N} \in M^N \mid \sum_{i \in N} \Delta^i = 0\}$, $w := (w^i)_{i \in N}$.

Proof. If $x \in M_{\mathcal{E}}(N)$, then $x = x(\mathcal{V}) = w + \Delta(\mathcal{V})$ for some $\mathcal{V} \in \mathcal{W}_M$, where $\Delta(\mathcal{V}) := (\Delta^i(\mathcal{V}))_{i \in N}$. By definition of M-contract it follows that $\Delta^i(\mathcal{V}) \in M$, $\sum_{i \in N} \Delta^i(\mathcal{V}) = 0$. So, consequently we obtain $x = w + \Delta$ for some $\Delta \in \Delta_M(N)$.

Let now $x = w + \Delta$ for some $\Delta \in \Delta_M(N)$. To prove that there is $\mathcal{V} \in \mathcal{W}_M$ such that $\Delta = \Delta(\mathcal{V})$, we apply induction on the number $|N|$ of the economical agents of $\mathcal{E}$. It is clear, that in case $|N| = 2$, due to the obvious inclusion $\{1, 2\} \in \Omega$, we have : $\mathcal{V} = \{\mathrm{v}\}$, with $\mathrm{v}^{12} := \Delta^1$, and $\mathrm{v}^{21} := \Delta^2$. Suppose that our assertion is valid for all economies of type (4.1) with $|N| \leq m$. Fix some economy $\mathcal{E}$ with $|N| = m + 1$. If $N \in \Omega$, then according to Remark 4.7, a modulo "zero-contract" c.s. as required consists of the unique M-contract v, defined by the formula: $\mathrm{v}^{ii+1} := \sum_{k=1}^{i} \Delta^k$, $i = 1, \dots, m$; $\mathrm{v}^{ij} = 0$, $j - i > 1$ (note that due to the equalities $\mathrm{v}^{ij} = -\mathrm{v}^{ji}$, to give a complete description of the contract v it is sufficient to indicate all the vectors v^{ij} with $i, j \in S(\mathrm{v})$ satisfying inequalities $i < j$).
Consider the case $N \notin \Omega$. Entering into the trivial contracts for some coalitions $S \in \Omega$, if necessary, we may assume without loss of generality,

that Ω is minimal (w.r.t. inclusion in 2^N) irreducible coalitional structure. Fix some $S^0 \in \Omega$. It is not hard to prove that there exists a partition $\eta^0 = \{N_1^0, N_2^0\}$ of N such that (i) coalitions $S_k^0 := S^0 \bigcap N_k^0$, $k = 1,2$, are nonempty; (ii) for any $T \in \Omega \setminus \{S^0\}$ either $T \subseteq N_1^0$, or $T \subseteq N_2^0$; and (iii) coalitional structures $\Omega_k^0 := \{T \in \Omega \mid T \subseteq N_k^0\} \bigcup \{S_k^0\}$, $k = 1,2$, are irreducible in N_1^0, N_2^0, respectively. Indeed, existence of a partition η^0, satisfying conditions (i) and (ii), follows immediately from the minimality of the irreducible coalitional structure Ω (since in case $\Omega \setminus \{T\}$ is irreducible for some $T \in \Omega$ we get a contradiction). As to the irreducibility of Ω_1^0, for example, it can easily be shown by considering three cases for any nontrivial partition $\{N_1^1, N_1^2\}$ of N_1^0: 1) $S^0 \bigcap N_1^k \neq \emptyset$, $k = 1,2$; 2) $S^0 \bigcap N_1^1 \neq \emptyset, S^0 \bigcap N_1^2 = \emptyset$; and 3) $S^0 \bigcap N_1^1 = \emptyset, S^0 \bigcap N_1^2 \neq \emptyset$. Specifically, considering partitions $\{N_1^2, N_1^1 \bigcup N_2\}$ and $\{N_1^1, N_1^2 \bigcup N_2\}$ in cases 2) and 3), respectively, we get the required coalitions from Ω_1^0 on the basis of condition (ii). In fact, since in both these cases coalition S^0 doesn't intersect one of the elements of the corresponding nontrivial partition of N, by (ii) and irreducibility of Ω, in each of these cases there exists a coalition $S \in \Omega_1^0$ such that $S \bigcap N_1^k \neq \emptyset$ for any $k = 1,2$.

Now, by making use of the partition η^0, we divide the initial economy $\mathcal{E}$ into two appropriate "smaller" economies

$$\mathcal{E}_{(k)} := \langle N_{(k)}, \{X_i^{(k)}, w_{(k)}^i, \alpha_i^{(k)}\}_{i \in N_{(k)}}, \Omega_{(k)} \rangle, \quad k = 1,2,$$

satisfying the inductive hypothesis and "implementing" the allocation $x = w + \Delta$ with $\Delta = (\Delta^1, \ldots, \Delta^n) \in \Delta_M(N)$ by means of some contractual M-system (as a result of a suitable exchange, both within and between these economies). To this end we fix some economic agents $i_k \in S_k^0$, $k = 1,2$, and put $N_{(k)} := N_k^0$, $k = 1,2$; $w_{(k)}^i := w^i$, $i \in N_{(k)}$, $k = 1,2$; and $X_i^{(k)} := X_i$ for any $i \in N_{(k)} \setminus \{i_k\}$, $k = 1,2$. As to the consumption sets $X_{i_1}^{(1)}$ and $X_{i_2}^{(2)}$, we put $X_{i_1}^{(1)} := \{w^{i_1} - \sum_{i \in N_{(1)} \setminus \{i_1\}} \Delta^i\}$ and $X_{i_2}^{(2)} := \{w^{i_2} + \sum_{i \in N_{(1)} \cup \{i_2\}} \Delta^i\}$. Finally, we take $\Omega_{(k)}$ to be equal to Ω_k^0, $k = 1,2$, and put $\alpha_i^{(k)} := \alpha_i$ for any $i \in N_{(k)}$ and $k = 1,2$.

To apply the inductive hypothesis, take $\Delta_{(k)} \in \Delta_M(N_{(k)})$, $k = 1,2$, with

$$\Delta_{(1)}^{i_1} := - \sum_{i \in N_{(1)} \setminus \{i_1\}} \Delta^i, \quad \Delta_{(2)}^{i_2} := \sum_{i \in N_{(1)} \cup \{i_2\}} \Delta^i,$$

and $\Delta_{(k)}^i := \Delta^i$ for any $i \in N_{(k)} \setminus \{i_k\}$, $k = 1,2$. Further, for each $k = 1,2$, put $w_{(k)} := (w^i)_{i \in N_{(k)}}$ and $\Delta_{(k)} := (\Delta_{(k)}^i)_{i \in N_{(k)}}$. It is clear, that

$x_{(1)} := w_{(1)} + \Delta_{(1)}$ and $x_{(2)} := w_{(2)} + \Delta_{(2)}$ are attainable allocations of the economies $\mathcal{E}_{(1)}$, $\mathcal{E}_{(2)}$, respectively. Hence, from this reasoning it follows that the economies $\mathcal{E}_{(1)}$, $\mathcal{E}_{(2)}$ and corresponding allocations $x_{(1)}$, $x_{(2)}$ satisfy our inductive hypothesis, and, consequently, there exist M-systems $\mathcal{V}_{(1)}$ and $\mathcal{V}_{(2)}$ such that $x_{(k)} = x(\mathcal{V}_{(k)}, \mathcal{E}_{(k)})$, each $k = 1, 2$.

Next we design a contract $\mathrm{v}_{(S^0)}$, which coalition S^0 should enter to in order to organize (together with the M-systems $\mathcal{V}_{(1)}$ and $\mathcal{V}_{(2)}$) an M-system $\mathcal{V}$, satisfying equality $x = x(\mathcal{V}) = x(\mathcal{V}, \mathcal{E})$. To do so, we put $\mathrm{v}^{i_1 i_2}_{(S^0)} := \sum_{i \in N_{(1)}} \Delta^i$, $\mathrm{v}^{i_2 i_1}_{(S^0)} := -\sum_{i \in N_{(1)}} \Delta^i$, and $\mathrm{v}^{ij}_{(S^0)} := 0$ for any $i, j \in S^0$ such that $\{i, j\} \neq \{i_1, i_2\}$. It is clear that the system $\mathcal{V} := \mathcal{V}_{(1)} \bigcup \mathcal{V}_{(2)} \bigcup \{\mathrm{v}_{(S^0)}\}$, with $\mathrm{v}_{(S^0)}$ thus defined (and modified in the spirit of Remark 4.7, if necessary), meets our requirement. This proves the proposition. □

To characterize the opportunities available to some coalition $S \in \Omega$ for to improve the allocations from $M_{\mathcal{E}}(N)$, we first describe the increments of the initial coalitional endowment $\sum_{i \in S} w^i$, generated by the breaking contracts belonging to some c.s. $\mathcal{V} \in \mathcal{W}_M$. Put

$$\mathcal{A}(\mathcal{V}, \mathcal{R}', S) := \bigcup_{\mathrm{w} | S(\mathrm{w}) = S} A(\mathcal{V}, \mathcal{R}', \mathcal{E}_{\mathrm{w}}), \quad \mathcal{A}(\mathcal{V}, S) := \bigcup_{\mathcal{R}' \subseteq \mathcal{R}^S} \mathcal{A}(\mathcal{V}, \mathcal{R}', S)$$

(here, according to Definition 4.4, we consider proper contracts w of type M only). Then all the increments, mentioned above, may be presented as the vectors $\Delta(\mathcal{V}', S) := \sum_{i \in S} \Delta^i(\mathcal{V}')$ with $\mathcal{V}' \in \mathcal{A}(\mathcal{V}, S)$. Note, that from the definition of $\mathcal{A}(\mathcal{V}, S)$ it follows that $x_S(\mathcal{V}')$ belongs to $\prod_{i \in S} X_i$ and, hence, $\sum_{i \in S} w^i + \Delta(\mathcal{V}', S) \in \sum_{i \in S} X_i$ for any $\mathcal{V}' \in \mathcal{A}(\mathcal{V}, S)$ (as usual, x_S is a restriction of $x = (x^1, \ldots, x^n)$ to S: $x_S = (x^i)_{i \in S}$ for any $S \subseteq N$).

Further, the set $\mathcal{X}_{\mathcal{V}}(S)$ of coalitional allocations, available to $S \in \Omega$ by means of breaking procedures applied to some M-system $\mathcal{V}$, is defined by

$$\mathcal{X}_{\mathcal{V}}(S) = \{z \in \prod_{i \in S} X_i^M \mid \exists \mathcal{V}' \in \mathcal{A}(\mathcal{V}, S) : z = w_S + \Delta_S(\mathcal{V}')\},$$

where $X_i^M := X_i \bigcap (w^i + M)$, $i \in N$, and, as above, $w_S = (w^i)_{i \in S}$, $\Delta(\mathcal{V}') = (\Delta^i(\mathcal{V}'))_{i \in N}$, and $\Delta_S(\mathcal{V}') = (\Delta^i(\mathcal{V}'))_{i \in S}$. To describe all the blocking possibilities of a coalition S at a state (allocation) $x \in M_{\mathcal{E}}(N)$, we have to take into account all the sets $\mathcal{X}_{\mathcal{V}}(S)$ with $\mathcal{V} \in \mathcal{W}(x)$. Observe, that the unions of these sets, $Z_x(S) := \bigcup_{\mathcal{V} \in \mathcal{W}(x)} \mathcal{X}_{\mathcal{V}}(S)$, are independent on the choice of c.a. x. Namely, the following assertion is valid.

Proposition 4.9 *For any $x, \widetilde{x} \in M_{\mathcal{E}}(N)$ and $S \in \Omega$ it holds $Z_x(S) = Z_{\widetilde{x}}(S)$.*

Proof. We fix some $S \in \Omega$, $x, \widetilde{x} \in M_{\mathcal{E}}(N)$ and demonstrate that $Z_x(S) \subseteq Z_{\widetilde{x}}(S)$. To do so, consider some non-empty contractual systems $\mathcal{V} = \{\mathrm{v}_r\}_{r\in\mathcal{R}} \in \mathcal{W}(x)$ and $\mathcal{V}_0 = \{\mathrm{u}_r\}_{r\in\mathcal{R}_0} \in A(\mathcal{V}, \mathcal{R}', \mathcal{E}_{\mathrm{w}}) \subseteq \mathcal{A}(\mathcal{V}, S)$, where w is an M-contract of the coalition S, and $\mathcal{R}'$ is some subset of $\mathcal{R}^S$. To show that there is a c.s. $\mathcal{V}' \in \mathcal{W}(\widetilde{x})$ yielding (after rescinding some suitable contracts from $\mathcal{V}'$) the same coalitional allocation for S, as $\mathcal{V}_0$ does, we consider first an arbitrary c.s. $\widetilde{\mathcal{V}} = \{\widetilde{\mathrm{v}}_r\}_{r\in\widetilde{\mathcal{R}}} \in \mathcal{W}(\widetilde{x})$. Since M is a linear subspace, satisfying the sign-assumption, we may assume w.l.g. (due to Remark 4.7), that $\widetilde{\mathcal{R}} = \mathcal{R}_0$ with $S(\widetilde{\mathrm{v}}_r) = S(\mathrm{u}_r)$ and $\widetilde{\mathrm{v}}_r^{ij} \neq \mathrm{u}_r^{ij}$ for any $r \in \mathcal{R}_0$ and $i, j \in S(\mathrm{u}_r)$ such that $i \neq j$. Put $\mathcal{V}' := \{\mathrm{v}_{(r,0)}\}_{r\in\mathcal{R}_0} \bigcup \{\mathrm{v}_{(r,1)}\}_{r\in\mathcal{R}_0}$, where $\mathrm{v}_{(r,0)}^{ij} := \mathrm{u}_r^{ij}$, $\mathrm{v}_{(r,1)}^{ij} := \widetilde{\mathrm{v}}_r^{ij} - \mathrm{u}_r^{ij}$. By construction of $\mathcal{V}'$ we get $\Delta(\mathcal{V}') = \Delta(\widetilde{\mathcal{V}})$. Hence, $\mathcal{V}'$ belongs to $\mathcal{W}(\tilde{x})$. Put $\mathcal{R} := \{(r, 1) \mid r \in \mathcal{R}_0^S\}$ and denote by $\mathcal{R}_1$ some maximal subset of $\mathcal{R}_0 \setminus \mathcal{R}_0^S$, satisfying that the c.s. $\widetilde{\mathcal{V}}_0 := \{\mathrm{v}_{(r,0)}\}_{r\in\mathcal{R}_0} \bigcup \{\mathrm{v}_{(r,1)}\}_{r\in\mathcal{R}_1}$ belongs to $A(\mathcal{V}', \mathcal{R}, \mathcal{E}_{\mathrm{w}})$ (note, that it can happen $\mathcal{R}_1$ to be empty set, since it may be necessary to dissolve all the contracts $\mathrm{v}_{(r,1)}$, $r \in \mathcal{R}_0 \setminus \mathcal{R}_0^S$, at the second step). Obviously, by definition of $\widetilde{\mathcal{V}}_0$ we have the equality $\Delta_S(\widetilde{\mathcal{V}}_0) = \Delta_S(\mathcal{V}_0)$. But this means that rescinding the contracts of c.s. $\mathcal{V}' \in \mathcal{W}(\widetilde{x})$ with the "names" from $\mathcal{R}$ makes it possible for the coalition S to get the same coalitional allocation, as the realization of $\mathcal{V}_0 \in A(\mathcal{V}, \mathcal{R}', \mathcal{E}_{\mathrm{w}})$ does at the state x.

Due to the arbitrariness of allocations $x, \widetilde{x} \in M_N(\mathcal{E})$, M-systems $\mathcal{V} \in \mathcal{W}(x)$, $\mathcal{V}_0 \in A(\mathcal{V}, \mathcal{R}', \mathcal{E}_{\mathrm{w}})$ with $\mathcal{R}' \subseteq \mathcal{R}^S$, and M-contract w, the argumentation given above proves the proposition. □

Since the set $Z_x(S)$ of coalitional allocations, accessible by coalition $S \in \Omega$ at the current state $x \in M_{\mathcal{E}}(N)$ as a result of breaking procedure, is independent on M-allocation x, it follows from Proposition 4.8 and obvious inclusion $x_S \in Z_x(S)$ that this set is determined completely by the characteristic function of cooperative game (in strategic form): $S \mapsto G_{\mathcal{E}}^M(S)$, $S \in \Omega \bigcup \{N\}$, where

$$G_{\mathcal{E}}^M(S) := Pr_S M_{\mathcal{E}}(N).$$

Above, as usual, $Pr_S M_{\mathcal{E}}(N)$ denotes a projection (restriction) of $M_{\mathcal{E}}(N)$ to $S:$

$$Pr_S M_{\mathcal{E}}(N) = \{x^S \in \prod_{i\in S} X_i \mid \exists x^{N\setminus S} \in \prod_{j\in N\setminus S} X_j \, [(x^S, x^{N\setminus S}) \in M_{\mathcal{E}}(N)]\}.$$

Definition 4.10 An allocation $x \in M_{\mathcal{E}}(N)$ is M-dominated by an allocation $y \in M_{\mathcal{E}}(N)$, if there exists a coalition $S \in \Omega$ such that $x^i \ \alpha_i \ y^i$ for every $i \in S$ and $y^i \in \mathcal{P}_i(x^i)$ for at least one $i \in S$.

The core of the M-domination relation, introduced above (i.e., the collection of all the allocations $x \in M_{\mathcal{E}}(N)$ that are not M-dominated by any allocation from $M_{\mathcal{E}}(N)$), is called the contractual M-core of the economy $\mathcal{E}$ and denoted by $C_0^M(\mathcal{E})$. From Propositions 4.8 and 4.9 and by definition of the contractual M-core and totally contractual set $D_0^M(\mathcal{E})$ of the economy $\mathcal{E}$ we obtain the next proposition.

Proposition 4.11 *For any exchange model $\mathcal{E}$ with irreducible coalitional structure Ω, and for any linear subspace M satisfying sign-assumption it holds that $D_0^M(\mathcal{E}) = C_0^M(\mathcal{E})$.*

4.2 Totally contractual set and equilibrium allocations

The reduction from $D_0^M(\mathcal{E})$ to $C_0^M(\mathcal{E})$ stated in Proposition 4.11 makes it possible to investigate the existence conditions, as well as to present an equilibrium characterization of the totally contractual allocations of a rather broad class of exchange models by means of conventional tools of cooperative game theory.

Below we restrict ourselves to the case that M is a hypersubspace of $\Re^l$ (i.e., we assume, that $\dim M = l - 1$). In this situation it turns out that any t.c. allocation is an equilibrium allocation under fairly natural regularity assumptions. Vice versa, under essentially weaker assumptions any equilibrium allocation turns out to be a t.c. allocation, provided that a subspace M is properly chosen. Recall, that $\bar{x} \in X(N)$ is called an equilibrium allocation, if there exists a nonzero vector $\bar{p} \in \Re^l$ such that

$$\mathcal{P}_i(\bar{x}^i) \bigcap B_i(\bar{p}) = \emptyset \tag{4.3}$$

for any $i \in N$, where $B_i(\bar{p}) := \{x^i \in X_i \mid \bar{p} \cdot x^i \leq \bar{p} \cdot w^i\}$ is the budget set of an agent i at prices $\bar{p}$. Denote by $W(\mathcal{E})$ the set of all equilibrium allocations of economy $\mathcal{E}$. As usual, any vector $\bar{p}$, satisfying (4.3), is called an equilibrium price vector, supporting the allocation $\bar{x}$.

Theorem 4.12 *Let $\bar{x}$ be an equilibrium allocation of the economy $\mathcal{E}$. If there is a strictly positive equilibrium price vector $\bar{p}$, supporting $\bar{x}$, then $\bar{x} \in D_0^M(\mathcal{E})$ with $M := \{z \in \Re^l \mid \bar{p} \cdot z = 0\}$.*

Proof. Since $\bar{x} \in W(\mathcal{E})$ implies the inclusion $\bar{x} \in X(N) \bigcap \prod_{i \in N} B_i(\bar{p})$, we have: $\Delta^i := \bar{x}^i - w^i$ belongs to M for any $i \in N$. It is clear, that irreducibility of Ω and Proposition 4.8 imply: $\bar{x} = w + \Delta \in M_{\mathcal{E}}(N)$, where, as before, $w = (w^i)_{i \in N}$, and $\Delta = (\Delta^i)_{i \in N}$. Suppose, the allocation $\bar{x}$ is M-dominated by some $x \in M_{\mathcal{E}}(N)$. Then, by definition of M-domination, there exist $S \in \Omega$ and $i \in S$ such that $x^i \in \mathcal{P}_i(\bar{x}^i)$. By definition of equilibrium allocation, the latter inclusion implies inequality $\bar{p} \cdot x^i > \bar{p} \cdot w^i$, which contradicts to the inclusion $x^i \in X_i^M$ yielding $\bar{p} \cdot x^i = \bar{p} \cdot w^i$.

Thus, $\bar{x} \in C_0^M(\mathcal{E})$, which means (according to Proposition 4.11) that the allocation $\bar{x}$ is a t.c. M-allocation of the exchange model $\mathcal{E}$. □

To prove the reverse inclusion $D_0^M(\mathcal{E}) \subseteq W(\mathcal{E})$ we have to add some traditional convexity and monotonicity assumptions, imposed on the consumption sets X_i and individual preference relations α_i (and some compatibility assumptions concerning the coalitional structure Ω). In what follows we suppose that for any $i \in N$ it holds: (i) X_i is a convex set; (ii) α_i is a reflexive binary relation; (iii) for any $x^i \in \widehat{X}_i := \mathrm{Pr}_{\{i\}}\, X(N)$ there exists $z \in \mathrm{int}\, \Re_+^l$ such that $x^i + z \in \mathcal{P}_i(x^i)$; (iv) for any $x^i \in X_i$ the set $\mathcal{P}_i(x^i)$ is convex and $(x^i, y^i] \subseteq \mathcal{P}_i(x^i)$ for any $y^i \in \mathcal{P}_i(x^i)$, where $(x^i, y^i] := \{(1-t)x^i + ty^i \mid t \in (0,1]\}$.

Following Vasil'ev (1984) a subset $T \subseteq N$ is said to be *Ω-divisible*, if for any $i \in N$ there is $S \in \Omega_i$ such that $T \bigcap (N \setminus S) \neq \emptyset$ (as before, $\Omega_i := \{S \in \Omega \mid i \in S\}$). Further, for any $x \in M_{\mathcal{E}}(N)$ put $N_x := \{i \in N \mid x^i \in \mathrm{int}_M X_i\}$, where $\mathrm{int}_M X_i$ is the relative interior of X_i^M in the affine subspace $w^i + M$. The main result of this subsection is as follows.

Theorem 4.13 *Suppose $\bar{x}$ belongs to $D_0^M(\mathcal{E})$ for some M satisfying sign-assumption. If $N_{\bar{x}}$ is an Ω-divisible subset of N, then $\bar{x}$ is an equilibrium allocation of $\mathcal{E}$.*

Proof. Let $\bar{x} \in D_0^M(\mathcal{E})$ and i be any agent of the economy $\mathcal{E}$. In order to show that $\bar{x}^i$ is a local maximum on X_i^M w.r.t. the binary relation α_i, we choose first some coalition $S_0 \in \Omega_i$ such that $k \notin S_0$ for some $k \in N_{\bar{x}}$ (recall that $N_{\bar{x}}$ is an Ω-divisible set). Since $\bar{x}^k \in \mathrm{int}_M X_k$, there is a symmetric neighborhood of zero in M, say U, such that $\bar{x}^k + U \subseteq X_k^M$. Suppose, there exists $z \in U$ satisfying inclusion $\bar{x}^i + z \in \mathcal{P}_i(\bar{x}^i)$. By Proposition 4.8, the allocation $y \in X(N)$, defined by the formula: $y^k := \bar{x}^k - z$, $y^i := \bar{x}^i + z$, $y^j := \bar{x}^j$ $(j \neq i, k)$, belongs to $M_{\mathcal{E}}(N)$. However, from the construction of y and definition of M-domination it follows that $\bar{x} \notin C_0^M(\mathcal{E})$. Indeed, since α_j is a reflexive binary relation for any $j \in N$, we have that the allocation y M-dominates $\bar{x}$ via coalition S_0. Because the latter contradicts Proposition 4.11, we get $\mathcal{P}_i(\bar{x}^i) \bigcap (\bar{x}^i + U) \bigcap X_i = \emptyset$. From this it follows that also the intersection $\mathcal{P}_i(\bar{x}^i) \bigcap X_i^M$ is empty. To show this, observe that for any $y^i \in \mathcal{P}_i(\bar{x}^i) \bigcap X_i^M$ (if such y^i exists) the bundle $\bar{z}(t) := ty^i + (1-t)\bar{x}^i$ belongs to the neighborhood $\bar{x}^i + U$ of $\bar{x}^i$ for any $t \in (0,1)$ small enough.

But this fact (together with the convexity assumption (iv)) contradicts the local maximality of $\bar{x}^i$, which has been proven above.

Summarizing, we have that for every $i \in N$ the consumption bundle $\bar{x}^i$ is maximal on X_i^M w.r.t. α_i. Fix now some $\bar{p} \in M^0 \bigcap \text{int } \Re^l_+$ and prove that $\mathcal{P}_i(\bar{x}^i) \bigcap B_i(\bar{p}) = \emptyset$ for any $i \in N$. Let $y^i \in \mathcal{P}_i(\bar{x}^i)$ for some $i \in N$. Since $\bar{p} \cdot x^i = \bar{p} \cdot w^i$ for any $x^i \in X_i^M$ (in particular, $\bar{p} \cdot \bar{x}^i = \bar{p} \cdot w^i$, each $i \in N$), from the maximality of $\bar{x}^i$ on X_i^M it follows that $\bar{p} \cdot y^i \neq \bar{p} \cdot w^i$. Suppose, that $\bar{p} \cdot y^i < \bar{p} \cdot w^i$. Because of the inclusion $\bar{x} \in X(N)$ we have, due to the monotonicity assumption (iii): there is $z \in \text{int } \Re^l_+$ such that $\bar{y}^i := \bar{x}^i + z \in \mathcal{P}_i(\bar{x}^i)$. It is clear, that $\bar{p} \cdot \bar{y}^i > \bar{p} \cdot w^i$. But then there exists $\bar{t} \in (0, 1)$ such that the bundle $z(\bar{t}) := \bar{t}y^i + (1 - \bar{t})\bar{y}^i$ belongs to X_i^M. On the other hand, due to the convexity of $\mathcal{P}_i(\bar{x}^i)$ we have: $z(\bar{t}) \in \mathcal{P}_i(\bar{x}^i)$, which contradicts the maximality of $\bar{x}^i$ on X_i^M w.r.t. α_i. From this we obtain the inequality $\bar{p} \cdot y^i > \bar{p} \cdot w^i$, which implies that $y^i \notin B_i(\bar{p})$. □

Note that the cardinality of $N_{\bar{x}}$ may be very small. It is clear, however, that the Ω-divisibility condition takes the simplest form in case $N_{\bar{x}} = N$.

Corollary 4.14 *If $\Omega \setminus \{N\}$ is a covering system (i.e., $\bigcup_{S \in \Omega | S \neq N} S = N$), then $D_{00}^M(\mathcal{E}) \subseteq W(\mathcal{E})$, where*

$$D_{00}^M(\mathcal{E}) := D_0^M(\mathcal{E}) \bigcap X_0^M, \; X_0^M := \prod_{i \in N} int_M X_i.$$

The results obtained allows us to find some rather large classes of exchange models, admitting an equilibrium characterization for any t.c. allocation $\bar{x}$, independently on the structure of the set $N_{\bar{x}}$. To give some examples, we introduce first some additional characteristics of the exchange model $\mathcal{E}$. We denote by $\mathcal{M}$ the set of all hypersubspaces $M \subseteq \Re^l$ satisfying the sign-assumption, and for any $M \in \mathcal{M}$ we put

$$\begin{aligned} S_{\mathcal{E}}^M \;\; := \;\; & \{i \in N \mid \alpha_i \text{ is complete, transitive, and} \\ & \alpha_i(w^i) \bigcap X_i^M \subseteq \text{int}_M X_i\}. \end{aligned}$$

Definition 4.15 An exchange model $\mathcal{E}$ is called C_M-regular, if $\alpha_i(w^i) \bigcap G_{\mathcal{E}}^M(\{i\}) \neq \emptyset$ for any $i \in S_{\mathcal{E}}^M$.

Definition 4.16 An exchange model $\mathcal{E}$ is called $C_{\mathcal{M}}$-regular, if it is C_M-regular for any $M \in \mathcal{M}$.

By applying, mutatis mutandis, the argumentation as used in the proof of Theorem 4.13 we obtain the following result.

Theorem 4.17 *Let $\mathcal{E}$ be C_M-regular exchange model. If $S_{\mathcal{E}}^M$ is Ω-divisible and $\{i\} \in \Omega$ for any $i \in S_{\mathcal{E}}^M$, then $D_0^M(\mathcal{E}) \subseteq W(\mathcal{E})$.*

Below we demonstrate some of the quite clear implications of Theorem 4.17.

Corollary 4.18 *If $\mathcal{E}$ is C_M-regular, $|S_{\mathcal{E}}^M| \geq 2$, and $\{i\} \in \Omega$ for any $i \in N$, then every totally contractual M-allocation is an equilibrium allocation.*

Corollary 4.19 *Let w^i belong to X_i for any $i \in N$. If $S_{\mathcal{E}}^M$ is Ω-divisible and $\{i\} \in \Omega$ for any $i \in S_{\mathcal{E}}^M$, then every totally contractual M-allocation is an equilibrium allocation.*

Corollary 4.20 *Let w^i belongs to X_i for any $i \in N$. If $|S_{\mathcal{E}}^M| \geq 2$, and $\{i\} \in \Omega$ for any $i \in N$, then every totally contractual M-allocation is an equilibrium allocation.*

Corollary 4.21 *If $\mathcal{E}$ is $C_{\mathcal{M}}$-regular, $S_{\mathcal{E}}^M$ is Ω-divisible for any $M \in \mathcal{M}$, and $\{i\} \in \Omega$ for every $i \in \bigcup_{M \in \mathcal{M}} S_{\mathcal{E}}^M$, then $D_0(\mathcal{E}) \subseteq W(\mathcal{E})$, where*

$$D_0(\mathcal{E}) := \bigcup_{M \in \mathcal{M}} D_0^M(\mathcal{E}).$$

To present a more easily verifiable regularity condition, similar to that given in Definition 4.16, we introduce another characteristic of the exchange model under consideration:

$$\begin{aligned} S_{\mathcal{E}} \;\; := \;\; & \{i \in N \mid \alpha_i \text{ is complete, transitive, and} \\ & \alpha_i(w^i) \bigcap \check{X}_i \subseteq \text{int } X_i\}, \end{aligned}$$

where $\check{X}_i := X_i \setminus (w^i + \text{int } \Re^l_+), \;\; i \in N$.

Definition 4.22 An exchange model $\mathcal{E}$ is called C-regular, if $\alpha_i(w^i) \bigcap G_{\mathcal{E}}^M(\{i\}) \neq \emptyset$ for any $i \in S_{\mathcal{E}}$ and $M \in \mathcal{M}$.

It is clear that a slight modification of the proof of Theorem 4.17 and Corollaries 4.18–4.21 yield the following analogs of Corollary 4.21.

Theorem 4.23 *If $\mathcal{E}$ is C-regular, $S_{\mathcal{E}}$ is Ω-divisible, and $\{i\} \in \Omega$ for every $i \in S_{\mathcal{E}}$, then $D_0(\mathcal{E}) \subseteq W(\mathcal{E})$.*

Corollary 4.24 *If $\mathcal{E}$ is C-regular, $|S_{\mathcal{E}}| \geq 2$, and $\{i\} \in \Omega$ for any $i \in N$, then $D_0(\mathcal{E}) \subseteq W(\mathcal{E})$.*

Corollary 4.25 *Let w^i belongs to X_i for any $i \in N$. If $S_{\mathcal{E}}$ is Ω-divisible and $\{i\} \in \Omega$ for any $i \in S_{\mathcal{E}}$, then $D_0(\mathcal{E}) \subseteq W(\mathcal{E})$.*

Corollary 4.26 *Let w^i belongs to X_i for any $i \in N$. If $|S_{\mathcal{E}}| \geq 2$, and $\{i\} \in \Omega$ for any $i \in N$, then $D_0(\mathcal{E}) \subseteq W(\mathcal{E})$.*

Observe that the assumptions of Corollary 4.26 are fulfilled for any exchange model $\mathcal{E}$ satisfying (i) w^i belongs to X_i for any $i \in N$, (ii) $\{i\} \in \Omega$ for any $i \in N$, and (iii) there are at least two agents with consumption sets X_i to be equal to $\Re^l_+$, and individual preferences α_i to be complete and transitive binary relations with $\alpha_i(w^i) \subseteq \text{int } \Re^l_+$.

As usual, we say that α_i is locally monotonic if for any $x^i \in \Pr_{\{i\}} X(N)$ there exists $\delta(x^i) > 0$ such that $x^i + z \in \mathcal{P}_i(x^i)$, whenever $z \in \Re^l_+$ and $0 < \|z\| < \delta(x^i)$. Taking into account that local monotonicity guarantees equilibrium prices to be strictly positive, and summarizing the results, which has been proven above, we obtain the following core-equivalence type theorem.

Theorem 4.27 *Suppose $\mathcal{E}$ satisfies the assumptions of either Theorem 4.23, or one of the Corollaries 4.21–4.26. If α_i is locally monotonic for at least one agent of the economy $\mathcal{E}$, then $D_0(\mathcal{E}) = W(\mathcal{E})$.*

Acknowledgement This chapter is the author's extended translation from Russian of the paper 'Characterization of the cores and generalized NM-solutions for some classes of cooperative games' originally published in *Transactions of Sobolev Institute of Mathematics*, 10, 1988. The translation was partly done while the author was visiting the Free University of Amsterdam, the Netherlands. Financial assistance from the Netherlands Organization for Scientific Research (NWO) in the framework of the Russian-Dutch programme for scientific cooperation, is gratefully acknowledged. The author would like to appreciate also financial support from the Russian Leading Scientific Schools Fund (grant 80.2003.6) and Russian Humanitarian Scientific Fund (grant 02-02-00189a).

References

1. Aliprantis, C.D., and K.C. Border (1994): *Infinite Dimensional Analysis.* Springer-Verlag, Berlin-Heidelberg-New York-Tokyo.

2. Aumann, R.J., and L.S. Shapley (1974): *Values of Nonatomic Games.* Princeton Univ. Press, Princeton, NJ.

3. Bondareva, O.N. (1962): "The theory of the core in an n-person game," *Vestnik Leningrad. Univ.*, 13, 141–142 (in Russian).

4. Dubey, P., A. Neyman, and R.J. Weber (1981): "Value theory without efficiency," *Math. Operation Research*, 6, 122–128.

5. Green, J.R. (1974): "Stability of Edgeworth's recontracting process," *Econometrica*, 42, 21–34.

6. Harsanyi, J.A. (1959): "A bargaining model for the cooperative n-person game," in: *Contributions to the Theory of Games IV* (Eds. A.W.Tucker and R.D.Luce), Princeton Univ. Press, Princeton, NJ, 325–355.

7. Kulakovskaya, T.E. (1971): "Necessary and sufficient conditions of coincidence of the core and NM-solution in classic cooperative game," *Soviet Math. Dokl.*, 199, n.5.

8. Lukas, V. (1971): "Some recent developments in an n-person game theory," *SIAM Review*, 13, 491–523.

9. Makarov, V.L. (1982): "Economical equilibrium: Existence and extremal properties," in: Problems of the Modern Mathematics, Moscow, *Nauka*, 19, 23–59 (in Russian).

10. Makarov, V.L., and V.A. Vasil'ev (1989): "On some problems and results of the modern mathematical economics," *Matekon*, 25, 4–95.

11. Maschler, M., and B. Peleg (1976): "Stable sets and stable points of set-valued dynamic systems with applications to game theory," *SIAM Journal of Control and Optimization*, 14, 985–995.

12. von Neumann, J., and O. Morgenstern (1953): *Theory of Games and Economic Behavior.* Princeton Univ. Press, Princeton, NJ.

13. Rockafellar, R.T. (1970): *Convex Analysis.* Princeton Univ. Press, Princeton, NJ.

14. Rosenmuller, J. (1981): *The Theory of Games and Markets.* North-Holland, Amsterdam.

15. Roth, A.E. (1976): "Subsolutions and supercore of cooperative games," *Mathematics of Operation Research*, 1, 43–49.

16. Shapley, L.S. (1971): "Cores of convex games," *International Journal of Game Theory*, 1, 11–26.

17. Vasil'ev, V.A. (1975): "The Shapley value for cooperative games of bounded polynomial variation," *Optimizatsiya*, 17, 5–27 (in Russian).

18. Vasil'ev, V.A. (1980): "On the H-payoffs in cooperative games," *Optimizatsiya*, 24, 18–32 (in Russian).

19. Vasil'ev, V.A. (1981a): "On a class of imputations in cooperative games," *Soviet Math. Dokl.*, 23, 53–57.

20. Vasil'ev, V.A. (1981b): "Lucas game has no NM-solution in H-payoffs," *Optimizatsiya*, 27, p.5–20 (in Russian).

21. Vasil'ev, V.A. (1982): "On a class of operators in a space of regular set functions," *Optimizatsiya*, 28, 102–111 (in Russian).

22. Vasil'ev, V.A. (1984): *Exchange Economies and Cooperative Games.* Novosibirsk State Univ. Press, Novisibirsk (in Russian).

23. Vasil'ev, V.A. (1987): "Generalized von Neumann-Morgenstern solutions and accessibility of cores," *Soviet Math. Dokl.*, 36, 374–378.

24. Vasil'ev, V.A. (1998): "The Shapley functional and polar forms of homogeneous polynomial games," *Siberian Adv. in Math.*, 8, 109–150.

25. Vasil'ev, V.A. (2003): "On extreme points of the Weber polyhedron," *Discrete Analysis and Operation Research*, Ser.1, 10, 17–55 (in Russian).

26. Vasil'ev, V.A., and M.G. Zuev (1988): "Support function of the core of a convex game on a metric compactum," *Optimizatsiya*, 44, 155–160 (in Russian).

Part III

Bargaining theory

Chapter 8

The linear bargaining solution

S. Pechersky

Abstract: The linearity, efficiency and symmetry axioms are used to define a linear bargaining solution. They lead to the "utilitarian rule", maximization of a weighted sum of criteria. The Steiner point of a convex set is used to define the linear solution of a bargaining game. Some modifications of the set of axioms are considered and the properties of solutions defined are studied.

Key words: Bargaining game, utilitarian rule, linear bargaining solution, Pareto optimality, Steiner point

1 Introduction

There are several theoretical and applied reasons for studying bargaining games (BG in what follows). Note only that many multi-criteria decision making problems can be reduced to BG. Besides, though BG are the simplest cooperative games (in the sense that the set of feasible utility vectors is nontrivial for the grand coalition only) they possess several important properties of general cooperative games.

J.Nash introduced two-person bargaining games in his seminal paper Nash (1950), and defined the Nash bargaining solution. Then many authors studied axiom systems defining different bargaining solutions and, in particular, existence and uniqueness of solutions. A detailed survey of different bargaining solutions is given in Roth (1979). Lately there is a growing interest in bargaining solutions based one way or another on addition of BG. So the superlinear n-person bargaining solution was, apparently, first defined in Pechersky (1979), where a corresponding existence theorem was proved. Perles and Maschler (1981) considered the superadditive bargaining solution for two-person bargaining games.

In this paper we study a set of axioms analogous to Shapley's set of axioms defining the Shapley value. It includes Linearity, Efficiency and Symmetry Axioms, and leads to the "utilitarian rule"—maximization of a weighted sum of criteria (cf., for example, Myerson, 1981; Thomson, 1981). We consider also some modifications of the set of axioms and study the properties of solutions defined.

2 Basic model

Definition 2.1 A n-person bargaining game is a pair $(\bar{q}, Q)$, where $\bar{q} \in I\!R^I$ is the *status quo* point, $Q \subset I\!R^I$ and $I = \{1, 2, \ldots, n\}$.

When interpreting this pair one can think as follows: if players act separately the only possible outcome for the players is $\bar{q}$ giving utility $\bar{q}_i$ to player $i = 1, 2, \ldots, n$. If players cooperate then they can potentially agree on any arbitrary outcome $q \in Q$.

Let G be a set of n-person bargaining games $(\bar{q}, Q)$. A *bargaining solution* (on G) is a function $f : G \to I\!R^n$ such that $f(\bar{q}, Q) \in Q$ for all $(\bar{q}, Q)$. The outcome $f(\bar{q}, Q)$ is the *solution of the bargaining game* $(\bar{q}, Q)$. Since the axiom of translation invariance for bargaining games is standard we fix the *status quo point* at $\mathbf{O} = (0, 0, \ldots, 0)$. This allows us to omit the status quo point in what follows and to identify a bargaining game $(\mathbf{O}, Q)$ with the set Q.

3 Hausdorff metric and sums

Let G_o^n be the set of all n-person bargaining games such that

(a) Q is compact and convex;

(b) there is $x > \mathbf{0}$ such that $x \in Q$.

Define on G_o^n a natural metric ρ setting

$$\rho(Q_1, Q_2) = \rho^H(Q_1, Q_2),$$

where ρ^H is the Hausdorff metric defined by the formula

$$\rho^H(Q_1, Q_2) = \inf\{\lambda \geq 0 : Q_1 + \lambda D \supset Q_2, Q_2 + \lambda D \supset Q_1\},$$

where D is the unit ball in $I\!R^n$. By continuity and convergence in what follows we understand the continuity and convergence in that metric.

Let us define addition of bargaining games in G_o^n:

$$Q_1 + Q_2 = \{x + y : x \in Q_1,\ y \in Q_2\}.$$

It is clear that $Q_1 + Q_2$ also possesses properties (a) and (b). This definition seems to be natural enough: if the utility vector x is feasible in Q_1 and y is feasible in Q_2 then in the sum of bargaining games the utility vector $x + y$ should be feasible.

Multiplication by a positive number λ is defined as usual: $\lambda Q = \{\lambda x : x \in Q\}$.

4 Linear solutions

Definition 4.1 A linear solution on G_o^n is a function $l : G_o^n \to I\!R^I$ such that following Axioms hold:

L1 (Weak Pareto optimality). For every $Q \in G_o^n$, $l(Q)$ is weakly Pareto optimal, i.e.

$$l(Q) \in \pi_o Q = \{x \in Q : \text{ there is no } y \in Q \text{ such that } y > x\}$$

.

L2 (Symmetry). Let τ be a permutation of the set $I = \{1, 2, \ldots, n\}$ then

$$l(\tau^* Q) = \tau^* l(Q),$$

where τ^* is the transformation of $I\!R^n$ induced by τ, i. e. $\tau^*(x) = (x_{\tau(1)}, \ldots, x_{\tau(n)})$.

L3 (Linearity). For $\lambda_1, \lambda_2 > 0$

$$l(\lambda_1 Q_1 + \lambda_2 Q_2) = \lambda_1 l(Q_1) + \lambda_2 l(Q_2).$$

The Linearity axiom can be interpreted as follows. Let a bargaining game R be a lottery of two bargaining games Q_1 and Q_2. Then the players "obeying the linear rule" are indifferent between reaching an agreement before or after the outcome of the lottery is available. Indeed, if $(\lambda, 1-\lambda)$, $0 < \lambda < 1$ defines a lottery, and the players reach an agreement immediately then their gains correspond to $l(\lambda Q_1 + (1-\lambda)Q_2)$. If they postpone the discussion to the moment after the lottery then the expectation of the postponed agreement is $\lambda l(Q_1) + (1-\lambda)l(Q_2)$.

Note that such an interpretation in the case of a "superlinear rule" says that the players prefer to reach an agreement before the outcome of the lottery is available (cf., for example, Perles and Maschler, 1981).

Let $v(Q,\cdot)$ be the support function of a set Q defined for any $\alpha \in I\!R^I$ by the formula

$$v(Q,\alpha) \equiv v_Q(\alpha) = \sup\{x\alpha : x \in Q\},$$

where $x\alpha = (x,\alpha) = x_1\alpha_1 + \cdots + x_n\alpha_n$ denotes the scalar product in $I\!R^n$. In what follows, we will consider the comprehensive hull of sets Q, so we restrict our attention to $\alpha \in I\!R^I_+$.

Proposition 4.2 *If l is a linear bargaining solution then $l(Q)e = v(Q,e)$ for every bargaining game $Q \in G^n_o$, where $e = (1,1,\ldots,1) \in I\!R^I$.*

Proof. Let us suppose to the contrary that there is a BG $Q_1 \in G^n_o$ such that

$$l(Q_1)e < v(Q_1,e). \tag{4.1}$$

Let $D_+ = D_n \cup I\!R^n_+$, where D^n is the unit ball in $I\!R^n$. By Axioms L1 and L2 we have

$$l(D_+) = \left(\frac{1}{\sqrt{n}},\ldots,\frac{1}{\sqrt{n}}\right) = \frac{e}{\sqrt{n}}.$$

Besides, $l(D_+)e = v(D_+,e)$ and

$$l(D_+)\alpha < v(D_+,\alpha)$$

for every $\alpha \in I\!R^I_+$ such that $\alpha \neq \lambda e$, $\lambda > 0$. Hence by L3 and (4.1) we have for every $\alpha \in I\!R^I$

$$l(D_+ + Q_1)\alpha = (l(D_+) + l(Q_1))\alpha < v(D_+,\alpha) + v(Q_1,\alpha) = v(D_+ + Q_1,\alpha)$$

which contradicts the efficiency of l. □

Thus a linear solution, if it exist, assigns to each bargaining game a point x maximizing the sum $x_1 + x_2 + \cdots + x_n$. It follows that actually the solution satisfies Pareto optimality:

$$l(Q) \in \pi Q = \{x \in Q : \text{ there is no } y \in Q,\ y \neq x \text{ such that } y \geq x\}.$$

Proposition 4.2 would be sufficient for the definition of a set-valued linear solution, but we are interested in a single-valued solution, and the existence of such solution will be proved in Theorem 6.1 below.

Proposition 4.3 *If l is a linear solution on G_o^n then it does not satisfy the Scale Transformation Invariance Axiom.*

Proof. Let us suppose the contrary and consider the BG defined by the set

$$Q = \mathrm{co}\{\mathbf{O}, (0,1), (2,0)\},$$

where co$A$ denotes the closed convex hull of a set A. By Proposition 4.2 $l(Q) = (2,0)$. Let now the first player's utilities be multiplied by $1/4$ and the second player's utilities be unchanged. Denote the corresponding BG by Q_1. Then the Scale Transformation Invariance Axiom implies $l(Q_1) = (1/2, 0)$, which contradicts Proposition 4.2. □

Proposition 4.4 *If l is a linear solution on G_o^n then it is not continuous.*

Proof. It is well known (see, for example, Rockafellar, 1970) that the subdifferential $\partial v(Q,\alpha)$ of the support function $v(Q,\cdot)$ at α is the contact set of Q and its support hyperplane with normal α. For bargaining games with single-valued $\partial v(Q,e)$ the solution $l(Q)$ is uniquely defined by Proposition 4.2.

Let $Q = \mathrm{co}\,\{\mathbf{O}, (1,0), (0,1)\}$. Define two sequences of BG $\{Q_m^1\}_{m=1}^{\infty}$ and $\{Q_m^2\}_{m=1}^{\infty}$, where

$$\begin{aligned} Q_m^1 &= \mathrm{co}\left\{\mathbf{O}, (1+\frac{1}{m}, 0), (0,1)\right\}, \\ Q_m^2 &= \mathrm{co}\left\{\mathbf{O}, (1,0), (0, 1+\frac{1}{m})\right\}. \end{aligned}$$

Clearly, Q_m^1 and Q_m^2 converge both to BG Q while $m \to \infty$.

By Proposition 4.2

$$\begin{aligned} l(Q_m^1) &= \left(1+\frac{1}{m}, 0\right), \\ l(Q_m^2) &= \left(0, 1+\frac{1}{m}\right). \end{aligned}$$

Let us suppose a linear solution to be continuous, i. e. $Q_m \to Q$ implies $l(Q_m) \to l(Q)$. Therefore letting $m \to \infty$ we set

$$l(Q) = \lim l(Q_m^1) = (1,0),$$

and

$$l(Q) = \lim l(Q_m^2) = (0,1),$$

which is impossible. □

5 Steiner point

In what follows we will use the Steiner point of a convex set. Let us recall the definition and some properties of the Steiner point (see, for example, Shephard, 1966; Meyer, 1970).

The Steiner point of an arbitrary closed bounded convex set K in $\mathbb{R}^n$ is defined by

$$s(K) = \frac{1}{\sigma_n} \int_{S^{n-1}} \alpha p(K, \alpha) d\omega,$$

where α is a variable unit vector, $p(K, \cdot)$ is the support function of K, ω is an element of the surface area of the unit sphere S^{n-1}, and σ_n is the volume of the n-dimensional unit ball. From the definition the linearity of $s(K)$, i.e.

$$s(\lambda_1 K_1 + \lambda_2 K_2) = \lambda_1 s(K_1) + \lambda_2 s(K_2),$$

and the continuity of $s(K)$ follow immediately.

The definition of the Steiner point can also be rewritten as follows (Shephard, 1966):

$$s(K) = \frac{1}{\tau_n} \int_{S^{n-1}} x(K, \alpha) d\omega,$$

where $\tau_n = n\sigma_n$ is the $(n-1)$-dimensional surface area of S^{n-1}, and $x(K, \alpha)$ is the contact set of the set K and support hyperplane $L(\alpha)$ with normal α, i.e.

$$x(K, \alpha) = L(\alpha) \cap K.$$

The Lebesgue integral

$$\int_{S^{n-1}} x(K, \alpha) d\omega$$

exists, since $x(K, \alpha)$ is a singleton except when α belongs to a set of measure zero on S^{n-1}.

Further, $s(K) \in K$, and if K is a singleton then $s(K) = K$.

The following theorem is due to Meyer.

Theorem (Meyer, 1970) *Let T be a set of orthogonal transformations of $\mathbb{R}^n$ such that*

(a) *T is transitive on the unit sphere S^{n-1}, i.e. if α_1 and $\alpha_2 \in S^{n-1}$, then there is $t \in T$ such that $t(\alpha_1) = \alpha_2$;*

(b) *for any $\alpha_0 \in S^{n-1}$ there exists a nonempty set $T(\alpha_0) \in T$ such that each t in $T(\alpha_0)$ fixes α_0, and if α is fixed by any t in $T(\alpha_0)$ then $\alpha = \lambda\alpha_0$ for some scalar λ.*

Let $f : \mathcal{K}^n \to I\!R^n$, where $\mathcal{K}^n$ is the set of all compact convex subsets of $I\!R^n$, be a linear, uniformly continuous function satisfying

$$f(tK) = tf(K)$$

for every $t \in T$ and $K \in \mathcal{K}^n$. Then $f(K) = \lambda s(K)$ for some fixed λ, where $s(K)$ is the Steiner point of the set K.

6 Existence of linear solutions

Now we prove the existence theorem.

Theorem 6.1 *There is a linear solution.*

Proof. Let Q_1 and Q_2 be two arbitrary bargaining games with set-valued sets $\partial v_{Q_i}(e)$, $i = 1, 2$. These sets are convex, compact and

$$\partial v_{Q_1+Q_2}(e) = \partial v_{Q_1}(e) + \partial v_{Q_2}(e). \tag{6.1}$$

The set $\partial v_{Q_1}(e)$ is a subset of an affine subspace $a_i + I\!R^{n-1}$ orthogonal to the vector e for some

$$a_i \in L_i(e) = \{x \in I\!R^n : xe = v_{Q_1}(e)\}.$$

Consider the set

$$\bar{V}_i = \partial v_{Q_i}(e) - a_i.$$

Clearly, $\bar{V}_i \in I\!R^{n-1}_e$, where $I\!R^{n-1}_e$ is the $(n-1)$-dimensional subspace of $I\!R^n$ orthogonal to e. Let $\bar{v}_i$ be its support function defined on the unit sphere $S^{n-2} \subset I\!R^{n-1}_e$. Define the Steiner point of a compact convex set $K \subset I\!R^{n-1}_e$ by the formula (cf. above)

$$s(K) = \frac{1}{\sigma_{n-1}} \int_{S^{m-2}} \alpha p(K, \alpha) d\omega, \tag{6.2}$$

where ω is the Lebesgue measure on the unit sphere $S^{n-2} \subset I\!R^{n-1}$, σ_{n-1} the volume of the unit ball in $I\!R^{n-1}$, $\alpha \in S^{n-2}$ and $p(K, \cdot)$ is the support function of a set K.

We have by (6.1)

$$\bar{V}_1 + \bar{V}_2 = \partial v_{Q_1+Q_2}(e) - a_1 - a_2 = \bar{V}.$$

The support function $\bar{v}$ of $\bar{V} \in I\!R^{n-1}$ is equal to $\bar{v}_1 + \bar{v}_2$. Hence we have by (6.2)

$$s(\bar{V}_1) + s(\bar{V}_2) = s(\bar{V}).$$

Let us prove now that $s(\bar{V}_1) + a_i$ does not depend on the choice of a point $a_i \in L_i(e)$. Indeed, let $b_i \in L_i(e)$, $b_i \neq a_i$, and consider the set $V_i^1 = \partial v_{Q_1}(e) - b_i$. Clearly

$$V_i = V_i^1 + b_i - a_i.$$

By linearity of the Steiner point we have $s(V_i) = s(V_i^1) + b_i - a_i$. Hence

$$s(V_i) + a_i = S(V_i^1) + b_i. \tag{6.3}$$

Let us define now

$$\bar{s}(\partial v_{Q_i}(e)) = s(V_i) + a_i. \tag{6.4}$$

By (6.3) this definition is correct.
Define a solution l by the following formula:

$$l(Q) = \bar{s}(\partial v_{Q_i}(e)). \tag{6.5}$$

Thus $l(Q)$ is defined as the Steiner point of a set $\partial v_Q(e)$ considered as $(n-1)$-dimensional subset of the corresponding hyperplane in $I\!R^I$. Such a solution satisfies Axiom L3 by linearity of the Steiner point and (6.1). Weak Pareto optimality follows from the definition. Check now the Symmetry. Let t' be a permutation of the set I. Then it induces an orthogonal transformation t of the subspace $I\!R_e^{n-1}$ orthogonal to e. For the support function $\bar{v}$ of a set in $I\!R_e^{n-1}$ we have

$$\begin{aligned}\bar{v}_{tQ}(\alpha) &= \bar{v}_{tQ}(t^{-1}(\alpha)) = \sup_{x \in tQ} xt(t^{-1}\alpha) = \sup_{y \in Q} ty \cdot t(t^{-1}\alpha) \\ &= \sup_{y \in Q} y \cdot t^{-1}\alpha = v_Q(t^{-1}\alpha).\end{aligned}$$

Since an orthogonal transformation preserves vector length, (6.4) implies Symmetry. □

Let $G_o' \subset G_o^n$ be the set of strictly convex bargaining games, and $G_o^1 \subset G_o^n$ be the set of such bargaining games Q that $\partial v(Q, e)$ is single-valued. Clearly $G_o' \subset G_o^1$.

Corollary 6.2 *A linear solution l is a unique linear solution on G_o^1.*

Corollary 6.3 *A linear solution l is a unique linear solution on G_o'.*

Proposition 6.4 *If $n = 2$ then l is a unique linear solution on G_o^n.*

Proof. In that case a solution $l(Q)$ necessarily coincides with the midpoint of the segment $\partial v(Q, e)$. Indeed, let us suppose the contrary. Without loss of generality we may assume that Q is a segment orthogonal to vector e or is a singleton (if not the proof remains almost the same). Our assumption means that there is a segment Q_1 which is orthogonal to e and nonsymmetric with respect to the bisector of the positive quadrant (otherwise, by L2, a solution should coincide with its midpoint) such that $l(Q_1) = (x_1, x_2) \neq (p_1, p_2)$, where (p_1, p_2) is the midpoint of Q_1.

Let $Q_2 = (p_2, p_1)$. Then, by L1, $l(p_2, p_1) = (p_2, p_1)$, and, by linearity

$$l(Q_1 + (p_2, p_1)) = (x_1 + p_2, x_2 + p_1). \tag{6.6}$$

However, $Q_1 + (p_2, p_1)$ is a segment orthogonal to e with the midpoint $(p_1 + p_2, p_2 + p_1)$. By symmetry, this is a solution of BG $Q_1 + (p_2, p_1)$, which contradicts 6.6. □

Definition 6.5 A linear solution is called weakly continuous in Q if $Q_m \to Q$ and $\partial v_{Q_m}(e) \to \partial v_Q(e)$ imply $l(Q_m) \to l(Q)$.

Definition 6.6 A linear solution is called weakly uniformly continuous if l is weakly uniformly continuous in every Q.

Proposition 6.7 *If $n = 3$ there is a unique weakly uniformly continuous linear solution.*

Proof. The existence of such a solution follows from Theorem 6.8. On the other hand, for $n = 3$ the dimension of $\partial v_Q(e)$ does not exceed 2. But in the two-dimensional case the Steiner point is the unique point satisfying continuity (in $\partial v_Q(e)$) and linearity (cf. Meyer, 1970). □

The following result may be interesting from a mathematical point of view.

Consider the following axiom.

L2'. Let t be an arbitrary orthogonal transformation of $\mathbb{R}^n$ preserving e. Then

$$l(tQ) = tl(Q).$$

Theorem 6.8 *There is a unique weakly uniformly continuous linear solution satisfying L1, L2' and L3.*

Proof. The linear solution defined by (6.5) satisfies the theorem, i.e., there is such solution.

By axiom L2' a solution l should be invariant with respect to any orthogonal transformation t of $I\!R^n$ preserving vector e. The invariant subspace of every such transformation is subspace $I\!R_e^{n-1}$ orthogonal to e. The group of all such transformations generates all orthogonal transformations of the subspace $I\!R_e^{n-1}$. By Proposition 4.2 $l(Q) = f(\partial v_Q(e))$ for some function f. By axiom L2' and the above-stated facts we have $\tau f(\partial v_Q(e)) = f(\tau \partial v_Q(e))$ for every orthogonal transformation of R_e^{n-1}, f being a linear and uniformly continuous function.

Thus by Meyer's Theorem (see above) $f(\partial v_Q(e)) = \lambda s(\partial v_Q(e))$ for some fixed λ. Let Q be such that $\partial v_Q(e) = x$ for some $x \in I\!R^I$, then $f(\partial v_Q(e)) = x = s(\partial v_Q(e))$. Hence $\lambda = 1$, and $l(Q) = s(\partial v_Q(l))$. Thus l so defined is the unique linear solution satisfying L1, L2', L3. □

7 Nonsymmetric linear solutions

Consider now our set of axioms without the symmetry axiom.

Definition 7.1 A function $l^N : G_o^n \to I\!R^I$ is called a nonsymmetric linear solution if it satisfies axioms L1 and L3.

Proposition 7.2 *If l^N is a nonsymmetric linear solution then there is an $\alpha \in I\!R_+^I$ such that $l^N(Q)\alpha = v_Q(\alpha)$ for every BG Q.*

Proof. Let $Q = D_+$ where $D_+ = D^n \cap I\!R_+^n$ and D_n is the unit ball in $I\!R^n$. Then $l^N(D_+) = x_0$ for some Pareto optimal point $x_0 \in D_+$. The unique support hyperplane at x_0 is defined by some normal $\alpha \in I\!R_+^n$.

Let $v_0(p)$ be the support function of D_+. Then $x_0\alpha = v_0(\alpha)$ and

$$x_0\beta < v_0(\beta) \quad \text{for} \quad \beta \neq \lambda\alpha,\ \lambda > 0 \tag{7.1}$$

Let now Q be an arbitrary BG and let us denote its support function by v. Let y be a solution of this BG, i.e. $l^N(Q) = y$. Then

$$y\beta \le v(\beta) \quad \text{for} \quad \beta \in I\!R_+^n.$$

By linearity $l^N(Q + D_+) = y + x_0$. Therefore

$$x_0\beta + y\beta \le v_0(\beta) + v(\beta).$$

By Pareto optimality there is a β such that

$$x_0\beta + y\beta = v_0(\beta) + v(\beta).$$

However, by (7.1) we have $x_0\beta + y\beta < v_0(\beta) + v(\beta)$ for any $\beta \neq \lambda\alpha$. Hence $x_0\alpha + y\alpha = v_0(\alpha) + v(\alpha)$. □

Theorem 7.3 *There is a nonsymmetric linear solution.*

The proof is similar to that of Theorem 6.1.

It follows immediately that there is a family of nonsymmetric linear bargaining solutions, each such solution being defined by some $\alpha \in \mathbb{R}^n_+$. Conversely, for every $\alpha \in \mathbb{R}^n_+$ there is a linear solution l^N_α. Clearly every nonsymmetric linear solution possesses all properties analogous to a symmetric linear solution.

Acknowledgement This chapter is a shortened version of Pechersky (1982). The author would like to express his gratitude to Hans Peters for his valuable comments and suggestions.

References

1. Meyer, W.J. (1970): "Characterization of the Steiner Point," *Pacific Journal of Mathematics*, 35, 717–725.

2. Myerson, R. (1981): "Utilitarianism, Egalitarianism and the Timing Effect in Social Choice Problem," *Econometrica*, 49, 883–897.

3. Nash, J.F. (1950): "The Bargaining Problem," *Econometrica*, 18, 155–162.

4. Pechersky, S. (1979): "On Bargaining Solutions," *Mathematical Methods in Social Sciences*, 88–107 (in Russian).

5. Pechersky, S. (1982): "Linear Bargaining Solution," *Mathematical Methods in Social Sciences*, 1982, 37–54 (in Russian).

6. Perles, M., and M. Maschler (1981): "The super-additive solution for the Nash bargaining game," *International Journal of Game Theory*, 10, 163–193.

7. Rockafellar, R.T. (1970): *Convex Analysis*. Princeton: Princeton University Press.

8. Roth, A.E. (1979): *Axiomatic Models of Bargaining*. Lecture Notes in Economic and Math. Systems, vol. 170, Springer Verlag.

9. Shepard, G. (1966): "The Steiner point of a convex polytope," *Canadian Journal of Mathematics*, 18, 1294–1300.

10. Thomson, W. (1981): "Nash's Bargaining Solution and Utilitarian Choice Rules," *Econometrica*, 49, 2, 535–538.

Chapter 9

On the superlinear bargaining solution

S. Pechersky

Abstract: A set of axioms is used to define a superlinear solution for n-person bargaining games. The existence of such a solution is proved.

Key words: Bargaining game, superlinear bargaining solution, partial additivity, Pareto optimality, Steiner point

1 Introduction

A n-person bargaining game is a pair $(\bar{q}, Q)$, where $\bar{q} \in \mathbb{R}^I$ is the *status quo* point, $Q \subset \mathbb{R}^I$ and $I = \{1, 2, \dots, n\}$. When interpreting this pair one can think as follows: if the players act separately the only possible outcome for the players is $\bar{q}$ giving utility $\bar{q}_i$ to player $i = 1, 2, \dots, n$. If the players cooperate then they can potentially agree on an arbitrary outcome $x \in Q$.

The axiomatic approach is most suitable to define a solution for bargaining games. There are several systems of axioms (cf., for example, Nash, 1950; Kalai and Smorodinsky, 1975; Butrim, 1978). We consider a new set of axioms, which contains a superlinearity axiom.

A bargaining game can be considered as a cooperative game without side payments where players forming any coalition, except the grand coalition, receive the gains corresponding to the *status quo* point. On the other hand games without side payments present a generalization of cooperative side payment games, which have been studied much more profoundly. This relates to different aspects and in particular to the axiomatic approach. It seems to be natural to aspire to eliminate this gap. The Shapley value (Shapley, 1953) is the most prominent solution concept for side payment games.

The most disputable among Shapley's axioms is the linearity axiom. There are several papers where the linearity axiom is replaced by some other axioms, the corresponding system of axioms being equivalent to that of Shapley. But there are no analogues to Shapley's linearity axiom.

We consider n-person bargaining games, and introduce a system of axioms reflecting an attempt to introduce an axiom analogous to the linearity axiom. The existence of a solution satisfying this set of axioms is proved.

2 Preliminaries

Let G be a set of n-person bargaining games $(\bar{q}, Q)$. A *bargaining solution (on G)* is a function $f : G \to \mathbb{R}^n$ such that $f(\bar{q}, Q) \in Q$. The outcome $f(\bar{q}, Q)$ is the *solution of the bargaining game* $(\bar{q}, Q)$. Since an axiom of translation invariance for bargaining games is standard we fix the *status quo* point at $\mathbf{O} = (0, 0, \ldots, 0)$. This allows us to omit the *status quo* point in what follows and to identify a bargaining game $(\mathbf{O}, Q)$ with the set Q.

Let G^n be the set of all n-person bargaining games such that

(a) Q is compact and convex;

(b) $x \geq \mathbf{O}$ for any $x \in Q$, i.e. we consider the individually rational payoff vectors only;

(c) there is $x > \mathbf{O}$ such that $x \in Q$.

Thus we identify a bargaining game with a compact convex set in $\mathbb{R}^n$.

The following notation is used throughout the paper:

$I = \{1, 2, \ldots, n\}$ denotes a set of players;

$T^{n-1} = \{x \in \mathbb{R}^n_+ : x_1 + \cdots + x_n = 1\}$ – the standard simplex;

$\text{comp}Q = \{x \in \mathbb{R}^n_+ : x \leq y \text{ for some } y \in Q\}$ – the comprehensive hull of a set Q;

$\text{cconv}A$ denotes the closed convex hull of a set Q;

$\pi Q = \{x \in Q : y \in Q, y \geq x \Rightarrow y = x\}$ the set of Pareto optimal points of a set Q;

$\pi_o Q = \{x \in Q : \text{there is no } y \in Q \text{ such that } y > x\}$ the set of weakly Pareto optimal points of a set Q.

Define on G^n a natural metric ρ setting

$$\rho(Q_1, Q_2) = \rho^H(Q_1, Q_2),$$

where ρ^H is the Hausdorff metric defined by formula

$$\rho^H(Q_1, Q_2) = \inf\{\lambda \geq 0 : Q_1 + \lambda D \supset Q_2,\ Q_2 + \lambda D \supset Q_1\},$$

where D is the unit ball in $I\!R^n$. By continuity and convergence in what follows we understand the continuity and convergence in that metric.

Let us define addition of bargaining games in G^n:

$$Q_1 + Q_2 = \{x + y : x \in Q_1, y \in Q_2\}.$$

It is clear that $Q_1 + Q_2$ also possesses properties (a)–(c). This definition seems to be natural enough: if utility vector x is feasible in Q_1 and y is feasible in Q_2 then in the sum of bargaining games utility vector $x + y$ should be feasible.

3 Axioms

Let $p : G^n \to I\!R^n$ be a bargaining solution. Impose the following axioms.

A1 (continuity). p is continuous in Q.

A2 (symmetry). Let τ be a permutation of the set $I = \{1, 2, \ldots, n\}$, then

$$p(\tau^* Q) = \tau^* p(Q),$$

where τ^* is the transformation of $I\!R^n$ induced by τ, i. e. $\tau^*(x) = (x_{\tau(1)}, \ldots, x_{\tau(n)})$.

A3 (weak Pareto optimality). $p(Q) \in \pi_0 Q$.

A4 (superlinearity). For λ_1, $\lambda_2 > 0$

$$p(\lambda_1 Q_1 + \lambda_2 Q_2) \geq \lambda_1 p(Q_1) + \lambda_2 p(Q_2).$$

A5 (partial additivity). If $q \in I\!R^n_{++}$ then

$$p(Q + q) = p(Q) + q.$$

Axioms A1, A2 and A3 need no comments. The efficiency axiom is formulated in a weakened form since, firstly, the set of Pareto optimal points does not depend continuously on Q (the corresponding map is lower semicontinuous), and, secondly, in accordance with Proposition 4.1 below there are "not many" bargaining games with $\pi_0 Q \neq \pi Q$.

Axiom A4 can be divided into two parts: positive homogeneity and super-additivity. Positive homogeneity is a well-known axiom of invariance under common changes of utility scales. The super-additivity axiom is a generalization of Shapley's additivity axiom to bargaining games: the players in BG-sum expect to receive at least as much as the sum of their gains in BG-components[1].

Axiom A5 is a refinement of axiom A4 for the case when one of the BG-components is a singleton. It can be interpreted as some kind of shift covariance axiom.

The bargaining solutions mentioned above (Nash, 1950; Kalai and Smorodinsky, 1975; Butrim, 1978) do not satisfy this set of axioms. Particular methods like equal distribution or distribution in accordance with given weights do not satisfy this system too (they do not satisfy A5).

We shall prove the existence of a superlinear solution. Unfortunately, it is not unique.

4 A superlinear solution

We start with an assertion concerning Pareto surfaces. We call a bargaining game Q non-levelled if $\pi_0(\text{cconv}\pi Q) = \pi Q$.

Proposition 4.1 *The subset of non-levelled bargaining games is dense in* G^n.

Proof.[2] Let Q be non Pareto optimal BG, i.e. $\pi_0 Q \neq \pi Q$. Clearly $\pi_0 Q \supset \pi Q$ for every BG Q. We shall construct a sequence Q_m of BG convergent to

[1] Now there is well-known interpretation of the superlinearity axiom: let a BG R be a lottery of bargaining games Q_1 and Q_2, then the players prefer to reach an agreement before the outcome of the lottery is available (see, for example, Perles and Maschler, 1981). Perles and Maschler in the mentioned paper proved that there exists a unique super-additive solution for two person bargaining games satisfying Pareto optimality, scale transformation invariance, super-additivity, symmetry and continuity. Their solution cannot be extended to n-person bargaining games for $n > 2$. Our solution is not scale transformation invariant but it is defined for an arbitrary n.

[2] This is a simplified version of the proof given in the original paper Pechersky (1979).

BG Q. Without loss of generality we may assume that Q is a comprehensive set, i. e.

$$y \leq x,\, y \geq \mathbf{O},\, x \in Q \Rightarrow y \in Q.$$

(Indeed, let us consider the set $Q' = \text{comp}Q$. Clearly $\pi_0(Q') \supset \pi_0(Q)$ and if Q' is not non-levelled then Q is not non-levelled, too (of course the converse is not true). Thus, if an increasing sequence of comprehensive bargaining games $\{Q'_k\}$ converges to a bargaining game Q' then the sequence $\{Q_k\}$ with $Q_k = Q'_k \cap Q$ converges to the bargaining game Q. Hence we can replace a set Q by its comprehensive hull $Q' = \text{comp}Q$).

Let $v(\cdot)$ be the support function of a comprehensive set Q defined for any $\alpha \in I\!R^I$ by the formula

$$v(\alpha) = \sup\{x\alpha : x \in Q\},$$

where $x\alpha = (x, \alpha) = x_1\alpha_1 + \cdots + x_n\alpha_n$ denotes the scalar product in $I\!R^n$. Since Q is comprehensive we restrict in what follows our attention to $\alpha \in I\!R^I_+$ only. (Note also that for every (weakly) Pareto optimal point $x \in Q$ there is an $\alpha \in I\!R^I_+$ such that $v(\alpha) = x\alpha$.)

Let Q be an arbitrary comprehensive bargaining game. Since Q is compact its support function $v(\alpha)$ is continuous (on $I\!R^I_+$). Let us consider a positive number $\lambda_1 < 1$. Then $\lambda_1 v(\alpha)$ is the support function of the comprehensive bargaining game $\lambda_1 Q$. Since $v(\alpha)$ is continuous there exist a natural number $m = m(\lambda_1)$ such that the comprehensive bargaining game Q_1 defined by

$$Q_1 = \{x \in I\!R^I_+ : x\alpha \leq \lambda_1 v(\alpha) \quad \text{for} \quad \alpha \in T^{n-1}_m\}, \tag{4.1}$$

where $T^{n-1}_m = \left\{\alpha \in T^{n-1} : \alpha_j \geq \frac{1}{m},\ i = 1, \ldots, m\right\}$, is contained in Q and its support function v_1 satisfies $v_1(\alpha) < v(\alpha)$ for any $\alpha \in I\!R^I_+$.

Let now λ_2 be such that $\lambda_1 < \lambda_2 < 1$ and $Q_1 \subset \lambda_2 Q$. We can repeat the previous procedure with λ_1 replaced by λ_2 and construct a set Q_2 (for some $m = m(\lambda_2) > m(\lambda_1)$) such that its support function v_2 satisfies $v_1(\alpha) < v_2(\alpha) < v(\alpha)$ for any $\alpha \in I\!R^I_+$. Then we continue the process recursively.

By construction the resulting sequence $Q_1, Q_2, \ldots$ possess the following properties. It converges to Q, and every bargaining game Q_k is non-levelled. (Nonlevelness follows from the fact that by construction (cf. (4.1)) for any $x \in \pi_0 Q_k$ there is an $\alpha > \mathbf{O}$ such that $x\alpha = v_k(\alpha)$). □

Theorem 4.2 *There exist a superlinear solution satisfying axioms A1–A5.*

Proof. We first construct a (non Pareto optimal) presolution and interpret it, and then this presolution will be extended to obtain the desired solution.

1) Let Q be a bargaining game in G^n with support function $v(\alpha)$. Below we restrict our attention to $\alpha \in I\!R^n_+$ and denote the corresponding restriction $v(\alpha)|_{I\!R^n_+}$ by $v(\alpha)$, too. Clearly, $v(\alpha)$ defines the comprehensive hull comp$Q$ of the set Q:

$$\text{comp}Q = \{x \in I\!R^n : x\alpha \leq v(\alpha), \text{ for all } \alpha \in I\!R^n_+\}.$$

Let us call a bargaining game $Q \in G^n$ smooth if it is non-levelled, the set πQ is smooth and for every $\alpha \in I\!R^n_+$, $\alpha \neq \mathbf{O}$ the contact set $x(\alpha)$ of a support hyperplane $L(\alpha)$ of the set Q with normal α is a singleton. (The last property means that πQ is a set of Pareto optimal points of some strictly convex set.)

Let Q be a smooth bargaining game. This means, in particular, that the gradient $\nabla v(\alpha)$ of its support function depends continuously on $\alpha \neq \mathbf{O}$. It is well-known that the contact set $x(\alpha)$ is a singleton and $x(\alpha) = \nabla v(\alpha)$.

Define a presolution $c(Q)$ of the bargaining game Q by the formula

$$c(Q) = \frac{1}{\Theta_{n-1}} \int_{T^{n-1}} x(\alpha) d\sigma, \tag{4.2}$$

where Θ_{n-1} is the $(n-1)$-dimensional surface area of simplex T^{n-1}, and $d\sigma$ is an element of surface area of T^{n-1}. (Note that the presolution $c(Q)$ is reminiscent of the Steiner point of a convex compact set (see, for example, Shepard, 1966).[3]

We can extend the definition to the nonsmooth case. Indeed, let Q_1, Q_2, ... be a sequence of smooth bargaining games convergent to a given BG Q. For every α the sequence $x^1(\alpha), x^2(\alpha), \dots$ of corresponding contact points is uniformly bounded. Hence there is a convergent subsequence and we may assume without loss of generality $x^k(\alpha)_{k\to+\infty} \to z$. Clearly $L^i(\alpha)$ converges to some support hyperplane $L(\alpha)$ of set Q, where $L^i(\alpha)$ is a support hyperplane of the set Q_i with normal α. Hence $z \in L(\alpha)$, and $z \in L(\alpha) \cap Q$ since $Q^k \to Q$. Besides, $L(\alpha) \cap Q$ is a singleton $x(\alpha)$ for every $\alpha \in T^{n-1}$ except for at most a null set in T^{n-1}. Thus the Lebesgue integral

$$\int_{T^{n-1}} x(\alpha) d\sigma$$

is defined, and by the bounded convergence theorem

$$\lim_{k\to\infty} \int_{T^{n-1}} x^k(\alpha) d\sigma = \int_{T^{n-1}} x(\alpha) d\sigma.$$

[3] See also S. Pechersky, The Linear bargaining solution (this volume).

Therefore, for every BG Q we define

$$c(Q) = \frac{1}{\Theta_{n-1}} \int_{T^{n-1}} x(\alpha) d\sigma. \tag{4.3}$$

Clearly $c(Q)$ is continuous in Q.

In the nonsmooth case the gradient transforms into the subdifferential $\partial v(\alpha)$ of v at α. Thus the presolution can be defined in general by the formula

$$c(Q) = \frac{1}{\Theta_{n-1}} \int \partial v(\alpha) d\sigma. \tag{4.4}$$

Note that $c(Q) = c(\text{comp}Q)$. It is well-known that $\partial v(\alpha)$ is a contact set of Q and the support hyperplane $L(\alpha)$, and, as mentioned above, it is a singleton for every $\alpha \in T^{n-1}$ except for at most a null set in T^{n-1}.

Let us consider a polyhedral bargaining game Q, and let v_j, $j = 1, \ldots, m$ be its vertices. Then $x(\alpha) = v_j$ for $\alpha \in T_j^{n-1}$, where $T_j^{n-1} = T^{n-1} \cap N^\circ(v_j)$, $N^\circ(v_j)$ being the interior of the normal cone of Q at v_j. Clearly the collection $\{T_j^{n-1}\}_{j=1}^m$ covers T^{n-1} except for a null set. Hence for a polyhedral set Q we have:

$$c(Q) = \sum_{j=1}^{m} v_j \psi(v_j), \tag{4.5}$$

where $\psi(v_j)$ is the ratio of the $(n-1)$-dimensional surface area of T_j^{n-1} to the surface area of T^{n-1}.

It follows, by construction, that $c(Q) \in \text{cconv}(\pi Q)$. We call it presolution since $c(Q) \notin \pi_0 Q$ except for flat sets $\pi_0 Q$.

2) Now consider the properties of the presolution defined above.

2.1. $c(Q)$ is additive, i.e.

$$c(Q_1) + c(Q_2) = c(Q_1 + Q_2).$$

This follows immediately from the fact that the support function of the sum of two sets equals the sum of their support functions.

2.2. $c(Q)$ is positively homogeneous, i.e.

$$c(\mu Q) = \mu c(Q) \quad \text{for} \quad \text{every} \quad \mu \geq 0.$$

This means that presolution is invariant under common rescaling of utilities.

The following properties are evident.

2.3. $c(\mu_1 Q_1 + \mu_2 Q_2) = \mu_1 c(Q_1) + \mu_2 c(Q_2)$ for any $\mu_1, \mu_2 \geq 0$.

2.4. $c(Q)$ is continuous in Q.

2.5. $c(Q) \in c\text{conv}(\pi Q)$

2.6. $c(Q)$ is individually rational, i.e. $c(Q) \geq \mathbf{O}$.

2.7. $c(Q)$ is weakly monotone in Q, i.e. if $Q_1 \subset Q_2$ and $x^1(\alpha) \leq x_2(\alpha)$ for every $\alpha \in I\!R^n_+$ then $c(Q_1) \leq c(Q_2)$.

2.8. $c(Q)$ is permutation invariant, i.e. if τ is a permutation of $I = \{1, \dots, n\}$ then $c(\tau^* Q) = \tau^* c(Q)$.

2.9. Let $q \in I\!R^n_+$ then $c(Q + q) = c(Q) + q$.

3) Let us now define a solution $p(Q)$. As we mentioned above the presolution of BG is Pareto optimal for flat sets πQ only. Taking into account the interpretation (see Remark 4.3 below) and the properties of presolution define a solution $p(Q)$ of a bargaining game Q as follows:

$$p(Q) = c(Q) + \mu e, \tag{4.6}$$

where $e = (1, 1, \dots, 1) \in I\!R^n$ and

$$\mu = \max\{\lambda : c(Q) + \lambda e \in Q\}. \tag{4.7}$$

Let us check superlinearity. Homogeneity is obvious since property 2.2 holds. Let now

$$\begin{aligned} p(Q_1) &= c(Q_1) + \mu_1 e, \\ p(Q_2) &= c(Q_2) + \mu_2 e, \\ p(Q_1 + Q_2) &= c(Q_1 + Q_2) + \mu_3 e. \end{aligned}$$

Since $c(Q_1 + Q_2) = c(Q_1) + c(Q_2)$ and $p(Q_1) + p(Q_2) \in Q_1 + Q_2$ it follows from (4.7) that $\mu_3 \geq \mu_1 + \mu_2$.

Hence $p(Q_1 + Q_2) - (p(Q_1) + p(Q_2)) = (\mu_3 - \mu_1 - \mu_2)e \geq \mathbf{O}$.

Now check partial additivity of p. Indeed, let

$$p(Q + q) = c(Q + q) + \mu e = c(Q) + q + \mu e,$$

and $p(Q) = c(Q) + \delta e$.

We have

$$\begin{aligned} \mu &= \max\{\lambda : c(Q + q) + \lambda e \in Q + q\} = \\ &= \max\{\lambda : c(Q) + q + \lambda e \in Q + q\} = \\ &= \max\{\lambda : c(Q) + \lambda e \in Q\} = \delta. \end{aligned}$$

Hence $p(Q + q) = p(Q) + q$. The remaining properties of solution p follow from the properties of presolution. □

Remark 4.3 An interpretation of the presolution is as follows. Let us suppose that the players' utilities are comparable with weights $\alpha_j \geq 0$, $i = 1, \ldots, n$. Then the maximal total utility available to the players is given by

$$\max_{x \in Q} \sum_{i=1}^{n} \alpha_i x_i = v(\alpha).$$

Thus $x(\alpha)$ (if singleton) represents the distribution of the maximal total utility, $\alpha_j, \ldots, \alpha_n$ being the utility comparision weights. Hence, the presolution $p(Q)$ can be interpreted as the vector of average utilities that players get when maximal total utility is distributed among the players, where the average is taken over the comparision weights $(\alpha_1, \ldots, \alpha_n) \in T^{n-1}$.

Remark 4.4 Note that the presolution is not invariant under linear rescaling of the players' utilities. Indeed, let

$$\begin{aligned} Q_1 &= \{x \in I\!R_+^{\{1,2\}} : x_1 + x_2 \leq 1\}, \\ Q_2 &= \{x \in I\!R_+^{\{1,2\}} : x_1 + 2x_2 \leq 2\}, \end{aligned}$$

i.e., $Q_2 = (2,1) \times Q_1$, where $\times$ denotes the coordinate-wise multiplication. Then, by (4.5), $c_2(Q_2) < c_2(Q_1)$.

Remark 4.5 It is not difficult to check that p does not satisfy the independence of irrelevant alternatives.

Remark 4.6 Unfortunately, nothing consoling can be said about the uniqueness of the superlinear solution, because, for example, the presolution defined clearly is not the unique additive function of Q (though it has a sufficiently transparent interpretation).

Acknowledgement This chapter is a shortened version of Pechersky (1979). The author would like to express his gratitude to Hans Peters for his valuable comments and suggestions.

References

1. Butrim, B. (1978): "N-person games with essential set of criteria," *Journal of Computational Mathematics and Mathematical Physics*, 18, N1, 62–72 (in Russian).

2. Kalai, E., and M. Smorodinsky (1975): "Other solutions to Nash's bargaining problem," *Econometrica*, 43, 513–518.

3. Nash, J.F. (1950): "The bargaining problem," *Econometrica*, 18, 155–162.

4. Pechersky, S. (1979): "One bargaining solution," *Mathematical Methods in Social Science*, 12, Vilnius, 88–107 (in Russian).

5. Perles, M., and M. Maschler (1981): "The super-additive solution for the Nash bargaining game," *International Journal of Game Theory*, 10, 163–193.

6. Shapley, L.S. (1953): "A value for n-person games," *Contributions to the theory of games*. Princeton, 2, 307–317.

7. Shepard, G. (1966): "The Steiner point of a convex polytope," *Canadian Journal of Mathematics*, 18, 1294–1300.

Chapter 10

Stable compromises under corrupt arbitration

N.S. Kukushkin

Abstract: An agent (arbiter) makes a decision determining payoffs to two other agents (players), each of whom promises a kickback as a function of the payment received. An equilibrium is defined and it is shown that there always exists an equilibrium which is Pareto better (for the players) than any other equilibrium.

Key words: Principal-agent, double agency, equilibrium

Let us consider the following situation. There are two players and there is a compact subset S of the plane with fixed coordinate axes, every $(x_1, x_2) \in S$ being interpreted as a vector of feasible payments (in transferable utility) to the players. An agent, called arbiter, is entitled to choose a point from S; the arbiter is free of any obligations and has no intrinsic preferences among $x \in S$. The players commit themselves to paying kickbacks to the arbiter contingent upon the payments received; then the arbiter chooses $x \in S$ maximizing his total payoff.

Remark 1 There is no need to understand this setting literally. For instance, the arbiter may be a country possessing valuable resources, but unable to process them herself; the players are then two international firms bidding for permission to exploit the resources. The assumption that the country is only interested in financial results is rather strong, but not unacceptable. Under another, a bit far-fetched, interpretation, the arbiter is an international body created by two countries for better coordination that has slid to serving its own interests instead.

Let us introduce necessary notation. We denote P the Pareto border of S, assuming for simplicity that $\min_{(x_1,x_2)\in P} x_i = 0$ ($i \in \{1,2\}$); $x_i^{\max}$ denotes $\max_{(x_1,x_2)\in P} x_i$. Strategies of each player i are functions f_i determining his payment to the arbiter dependent on his own payment, $f_i(x_i)$. The functions f_i are assumed defined and nonnegative for all $x_i \in [0, x_i^{\max}]$; it would seem natural to assume $f_i(x_i) \le x_i$, but this inequality is automatically satisfied at an equilibrium to be defined, because each player is able to make his net payoff, $x_i - f_i(x_i)$, nonnegative. Without restricting generality, the functions f_i may be assumed nondecreasing; then the arbiter can restrict himself to outcomes $(x_1, x_2) \in P$. We impose the assumptions.

Definition 2 An equilibrium is a quadruple $(x_1^0, x_2^0, f_1^0, f_2^0)$ such that:

1. $(x_1^0, x_2^0) \in P$ and for every $(x_1, x_2) \in P$, the inequality $f_1^0(x_1^0) + f_2^0(x_2^0) \ge f_1^0(x_1) + f_2^0(x_2)$ holds;
2. $x_1^0 - f_1^0(x_1^0) \ge x_1^* - f_1^*(x_1^*)$ for every f_1^* and every $(x_1^*, x_2^*) \in P$ which maximizes $f_1^*(x_1) + f_2^0(x_2)$ over P;
3. $x_2^0 - f_2^0(x_2^0) \ge x_2^* - f_2^*(x_2^*)$ for every f_2^* and every $(x_1^*, x_2^*) \in P$ which maximizes $f_1^0(x_1) + f_2^*(x_2)$ over P.

Proposition 3 *Let $(x_1^0, x_2^0) \in P$, $y_1^0 \le x_1^0$, and $y_2^0 \le x_2^0$. The existence of functions f_1^0 and f_2^0 such that $(x_1^0, x_2^0, f_1^0, f_2^0)$ is an equilibrium and $f_i^0(x_i^0) = y_i^0$ for both i is equivalent to this system of inequalities:*

$$\begin{array}{rcl} y_1^0 + y_2^0 & \le & \min[x_1^{\max}, x_2^{\max}], \\ y_1^0 \ge x_2^{\max} - x_2^0, & & y_2^0 \ge x_1^{\max} - x_1^0. \end{array} \tag{1}$$

Proof. Let inequalities (1) hold. We define the functions f_1^0 and f_2^0 in this way:

$$\begin{aligned} f_i^0(x_i) &= 0 \text{ for } x_i < x_i^0; \\ f_i^0(x_i) &= y_i^0 \text{ for } x_i^0 \le x_i < x_i^{\max}; \\ f_i^0(x_i^{\max}) &= y_1^0 + y_2^0. \end{aligned}$$

If, e.g., $x_1^0 = x_1^{\max}$, then our general assumptions imply $y_2^0 = x_2^0 = 0$, so the definition of f_i^0 is consistent.

If both players have chosen their f_i^0, the arbiter, obviously, cannot get more than $y_1^0 + y_2^0$ whereas he gets as much by choosing (x_1^0, x_2^0), so the first condition holds.

Let us check the second condition. Let f_i^* be an arbitrary strategy of player 1, and $(x_1^*, x_2^*) \in P$ be such that

$$x_1^* - f_1^*(x_1^*) > x_1^0 - f_1^0(x_1^0) = x_1^0 - y_1^0. \tag{2}$$

If $x_1^* = 0$, the strict inequality in (2) is impossible. If $0 < x_1^* < x_1^0$, then (2) implies $f_1^*(x_1^*) < y_1^0$; however, $f_2^0(x_2^*) = y_2^0$, hence $f_1^*(x_1^*) + f_2^0(x_2^*) < y_1^0 + y_2^0 \leq f_1^*(0) + f_2^0(x_2^{\max})$, i.e., the arbiter would prefer $(0, x_2^{\max}) \in P$ to (x_1^*, x_2^*). Finally, if $x_1^* > x_1^0$, then $x_2^* < x_2^0$ and $f_2^0(x_2^*) = 0$, wherefrom, taking into account (2), we have $f_1^*(x_1^*) + f_2^0(x_2^*) < x_1^* - x_1^0 + y_1^0$; now (1) implies $f_1^*(x_1^*) + f_2^0(x_2^*) < y_1^0 + y_2^0 \leq f_1^*(0) + f_2^0(x_2^{\max})$ and the arbiter again prefers $(0, x_2^{\max})$.

The third condition is checked in the same way.

Now let us prove the necessity of (1). Let $(x_1^0, x_2^0, f_1^0, f_2^0)$ be an equilibrium and $y_i^0 = f_i^0(x_i^0)$ $(i \in \{1, 2\})$. Suppose that the first inequality in (1) does not hold, i.e., $y_1^0 + y_2^0 = \min[x_1^{\max}, x_2^{\max}] + 2d$ with $d > 0$. Without restricting generality, $x_2^{\max} \leq x_1^{\max}$; we define f_1^* in this way:

$$\begin{aligned} f_1^*(x_1) &= 0 \text{ for } x_1 < x_1^0; \\ f_1^*(x_1) &= f_1^0(x_1) - d \text{ for } x_1 \geq x_1^0. \end{aligned}$$

Let player 1 use f_1^*, player 2 use f_1^0, and $(x_1^*, x_2^*) \in P$ maximize the arbiter's payoff, $f_1^*(x_1) + f_2^0(x_2)$. Choosing (x_1^0, x_2^0), the arbiter receives $y_1^0 + y_2^0 - d > x_2^{\max}$; choosing $x_1^* < x_1^0$, he cannot receive more than $x_2^{\max}$. Therefore, $x_1^* \geq x_1^0$, whereas player 1 pays $f_1^0(x_1) - d$, hence $x_1^* - f_1^*(x_1^*) > x_1^0 - f_1^0(x_1^0)$, contradicting the second condition in the definition of equilibrium.

If $y_1^0 + y_2^0 \leq \min[x_1^{\max}, x_2^{\max}]$, but $x_1^{\max} - x_1^0 = y_2^0 + 2d$ with $d > 0$, we define f_1^*:

$$\begin{aligned} f_1^*(x_1) &= f_1^0(x_1) \text{ for } 0 \leq x_1 < x_1^{\max}; \\ f_1^*(x_1) &= y_1^0 + y_2^0 + d \text{ for } x_1 = x_1^{\max}. \end{aligned}$$

Let player 1 use f_1^* and player 2 use f_1^0. If the arbiter chooses $(x_1, x_2) \in P$ with $x_1 < x_1^{\max}$, he receives $f_1^*(x_1) + f_2^0(x_2) = f_1^0(x_1) + f_2^0(x_2) \leq f_1^0(x_1^0) + f_2^0(x_2^0) = y_1^0 + y_2^0 < y_1^0 + y_2^0 + d \leq f_1^*(x_1^{\max}) + f_2^0(0)$; therefore, $(x_1^*, x_2^*) = (x_1^{\max}, 0)$ will be chosen, giving to player 1 net payoff $x_1^{\max} - y_1^0 - y_2^0 - d = x_1^0 - y_1^0 + d > x_1^0 - y_1^0$, which contradicts the second condition in the definition of an equilibrium.

The necessity of the second inequality is checked in the same way. □

Let $(x_1^{00}, x_2^{00}) \in P$ be such that

$$x_1^{00} + x_2^{00} = \max_{(x_1, x_2) \in P} [x_1 + x_2];$$

denoting $y_1^{00} = x_2^{\max} - x_2^{00}$ and $y_2^{00} = x_1^{\max} - x_1^{00}$, we easily see that $y_i^{00} \le x_i^{00}$ for both i, and the inequalities (1) hold for $(x_1^{00}, x_2^{00}, y_1^{00}, y_2^{00})$. Therefore, there always exist equilibria as defined above.

Proposition 4 *Let* $(x_1^0, x_2^0, f_1^0, f_2^0)$ *be an equilibrium; then*

$$x_1^0 - f_1^0(x_1^0) \le x_1^{00} - y_1^{00} = \max_{(x_1,x_2)\in P} [x_1 + x_2] - x_2^{\max}$$

and

$$x_2^0 - f_2^0(x_2^0) \le x_2^{00} - y_2^{00} = \max_{(x_1,x_2)\in P} [x_1 + x_2] - x_1^{\max}.$$

Moreover, an equality (anywhere) is only possible when $x_1^0 + x_2^0 = x_1^{00} + x_2^{00}$.

Proof. We only have to rearrange the inequalities (1), denoting $y_i^0 = f_i^0(x_i^0)$. We have $x_1^0 \ge y_1^0 \ge x_2^{\max} - x_2^0$ and $x_2^0 \ge y_2^0 \ge x_1^{\max} - x_1^0$, hence $x_1^0 - y_1^0 \le x_1^0 + x_2^0 - x_2^{\max}$ and $x_2^0 - y_2^0 \le x_1^0 + x_2^0 - x_1^{\max}$. □

Proposition 4 effectively means that the net payoffs of both players attain a common maximum over the set of all equilibria. It is also worth noting that $y_1^0 + y_2^0 \ge x_1^{\max} + x_2^{\max} - (x_1^0 + x_2^0)$, i.e., the arbiter gets the least at any equilibrium best for the players (it need not be unique).

Note to the translation In more modern terms, this paper is about double agency. However, the formulation of the problem is different from Bernheim and Whinston (1986), and the results seem independent.

Acknowledgement Translated from Russian; originally published in: Matematicheskie Modeli Povedeniya. Vypusk 4. Saratov, 1978, pp. 85–88.

References

1. Bernheim, B.D., and M.D. Whinston (1986): "Common Agency," *Econometrica*, 54, 923–942.

Part IV

Equilibrium theory

Chapter 11

Stability of economic equilibrium

F.L. Zak

Abstract: We consider a standard Walrasian pure exchange economy. From an economic point of view, equilibrium in such a model only makes sense if none of the participants can benefit from throwing out some of the commodities. Equilibria with this property are called *stable*. Examples show that an equilibrium need not be stable even if it is unique and the preferences are defined by nice utility functions. In this paper we study stability from an infinitesimal point of view and give sufficient conditions for (local and global) stability.

Key words: Walrasian economy, equilibrium, stability, throwing out of commodities

1 Introduction

The classical problem of existence of equilibrium prices in competitive economy (i.e. prices for which demand equals supply) was solved in the fifties in the classical paper by Arrow and Debreu (cf. Intriligator, 1971; Nikaido, 1997). By now the results of these authors have been considerably generalized. However, the models considered in most of the papers on economic equilibrium are just modifications of the Walrasian model introduced more than a century ago. In this model, the producers' goal is to maximize their incomes and the consumers' goal is to maximize their utilities.

The Walrasian model has been repeatedly criticized in the economic literature, but only recently it has been noticed that even the restrictions imposed on the consumers' behavior are not natural within the framework of the model. Inspired by the classical result of Samuelson on the advantages provided by foreign trade, Gale (1974) constructed an example of a pure exchange economy in which some agents can redistribute their initial endowments between themselves in such a way that the new equilibrium state is better for each of them than the old one (of course, some of the outsiders are bound to suffer). However, it should be noted that cooperation between economic agents contradicts the assumptions of independence of economic agents and competitiveness of Walrasian economy.

Aumann and Peleg (1974) gave an example of a pure exchange economy in which some of the participants can improve their position (in equilibrium) by throwing out (or hiding) some of their initial endowments. Economies in which this is possible are called *unstable*, and those in which this is impossible are called *stable*. It is clear that the Walrasian approach makes sense only with respect to stable economies. At the same time, unstable economies not only naturally arise in economic theory, but also reflect certain economic realities.

Examples of this phenomenon are given by overproduction crises when, in order to keep prices high, producers destroy surplus commodities, and by "humanitarian" aid when commodities are transferred to outsiders whose participation in trade is insignificant. Besides these extreme examples, there are numerous others when producers benefit from cutting down production in order to improve their position by inflating prices.

It therefore seems to be desirable to develop a mathematical theory of stable and unstable economies. Recently, first steps on the way of developing such a theory were made by several authors (cf. e.g. Efimov and Shapovalov, 1979; Polterovich and Spivak, 1978). In the present paper we consider this problem from the point of view of differentiable manifolds.

We consider the following economic setup. There are m economic agents with demand functions $f_1(p, w_1), \dots, f_m(p, w_m)$ (where p is the price vector and w_k is the income of the k-th participant) with initial endowments $\omega_1, \dots, \omega_m$ (where ω_k is a nonnegative l-vector and l is the number of commodities). In this paper we consider the case that the equilibrium price vector is unique (modulo a nonzero scalar factor). However our methods also allow to treat the general case as well.

Let $p(\omega)$ be the equilibrium price vector corresponding to initial endowments ω. Then, as a result of trade, the k-th participant acquires a commodity vector $f_k(p(\omega), \langle \omega_k, p(\omega) \rangle)$ $(k = 1, \dots, m)$. Suppose that the

i-th participant threw out a part of his initial endowment, and let $\omega' = \{\omega_1, \ldots, \omega_i', \ldots, \omega_m\}$, $\omega_i' \leq \omega_i$ be the new initial endowment. Then stability or instability of ω is determined by the relation between the utilities of $f_i(p(\omega), \langle \omega_i, p(\omega) \rangle)$ and $f_i(p(\omega'), \langle \omega_i', p(\omega') \rangle)$. The present paper is devoted to a study of the behavior of these utilities under variation of initial endowments.

Throwing out commodities does not always seem realistic. Often it is more natural to assume that commodities are hidden and being stored. However, the lack of good dynamic models describing several cycles of production and consumption forces us to limit ourselves to the case of throwing out. From the point of view of economic theory this can be justified by assuming that we consider only perishable commodities (or commodities whose storage is very costly).

We proceed with briefly describing the organization of the paper. In the first section we generalize results of Balasko (1975), Debreu (1970), and Dierker (1974) and give a characterization of regular economies convenient for our purposes. In the second section we introduce the notion of infinitesimal stability and prove several criteria for infinitesimal stability. Using these criteria, we prove infinitesimal stability of economies with normal demand and equilibrium gross substitutability as well as economies close to equilibrium. Special attention is devoted to boundary economies (i.e. to the case when not all of the economic agents have all the commodities present in their initial endowments) since this case often occurs in examples and applications. In the third section we prove the main economic results of this paper. In particular, it is shown that in the case of normal demand the economic agents can only lose on discarding (positive amounts of) all the commodities (cf. 4.7) and that economies close to equilibrium are stable (cf. 4.6). It is also shown that generic infinitesimal stability implies stability (cf. 4.4, 4.9 and 4.10), and that economies with normal demand and equilibrium gross substitutability are stable (cf. 4.11).

It should be noted that our Theorem 4.11 overlaps with the main theorem 11 from Polterovich and Spivak (1978), although none of these theorems follows from the other one. Furthermore, Polterovich and Spivak study the notion of *coalition stability* (commodities can be redistributed between members of some coalition; cf. also Efimov and Shapolov, 1979). Our methods also allow to consider coalition stability.

2 Regular economies

2.1 We consider a pure exchange market with m economic agents and l commodities. Denote by $\mathbb{R}^k_+$ (resp. $\overline{\mathbb{R}}^k_+$) the open positive (resp. closed nonnegative) orthant of the k-dimensional Euclidean vector space $\mathbb{R}^k$, and let $f_i(p, w_i)$, where $p \in \mathbb{R}^l_+$ is a price vector and $w_i \in \mathbb{R}^1_+$ is the income of the i-th participant ($i = 1, \ldots, m$), be the demand function of the i-th participant. [1]

We assume that $f_i : \mathbb{R}^l_+ \times \mathbb{R}^1_+ \to \mathbb{R}^l_+$, $f_i(\mathbb{R}^l_+ \times \{0\}) = 0$ and f_i is generated by a preference relation on $\overline{\mathbb{R}}^l_+$ defined by a strictly monotone strictly quasiconcave twice continuously differentiable utility function $u_i : \overline{\mathbb{R}}^l_+ \to \mathbb{R}^1_+$:

$$f_i(p, w_i) = \{x \mid u_i(x) = \max u_i(y),\ y \in \overline{\mathbb{R}}^l_+, \langle y,\ p\rangle = w_i\}. \tag{2.1}$$

Thus f_i is a well defined (single valued) continuously differentiable (cf. Debreu, 1972) function satisfying the Walras relation

$$\langle p, f_i(p, w_i)\rangle = w_i, \quad i = 1, \ldots, m. \tag{2.2}$$

Furthermore, from (2.1) it follows that f_i is homogeneous of degree zero, i.e. demand does not change under price rescaling:

$$f_i(\alpha p, \alpha w_i) = f_i(p, w_i), \quad \alpha \in \mathbb{R}^l_+,\ i = 1, \ldots, m. \tag{2.3}$$

2.2 Let $f : \mathbb{R}^l_+ \times (\overline{\mathbb{R}}^1_+)^m \to \overline{\mathbb{R}}^l_+$,

$$f(p; w_1, \ldots, w_m) = \sum_{i=1}^{m} f_i(p, w_i) \tag{2.4}$$

be the aggregate demand function of all the economic agents. In what follows we need to impose a certain condition on the behavior of f near the boundary of the positive price orthant. Here follows such a condition which seems most convenient for our purposes.

$$\begin{array}{l} \text{Let } p_n \in \mathbb{R}^l_+,\ w_n = ((w_1)_n, \ldots, (w_m)_n) \in \mathbb{R}^m_+,\ n = 1, 2, \ldots; \\ p_n \to \bar{p} \in (\partial \overline{\mathbb{R}}^l_+ \setminus \{0\}),\ w_n \to \overline{w} \in \mathbb{R}^m_+ \setminus \{0\}. \\ \text{Then } \| f(p_n; w_n) \| \to \infty. \end{array} \tag{A}$$

[1] Throughout the paper subscripts refer to agents, whereas superscripts refer to components, e.g. f_i^j is the demand of consumer i for commodity j.

2.3 Let $\omega_i \in \overline{I\!R}^l_+$ be the vector of initial endowments of the i-th participant $(i = 1, \ldots, m)$. For given demand functions, the economy is defined by the initial endowments, thus the collection of all economies is

$$\overline{\Omega} = \{\omega \mid \omega_i \geq 0, \quad i = 1, \ldots, m; \sum_{i=1}^{m} \omega_i > 0\}. \tag{2.5}$$

Notice that we assume that for any economy $\omega = (\omega_1, \ldots, \omega_m) \in \overline{\Omega}$ all the commodities are represented in the market.

We say that $p \in I\!R^l_+$ is an *equilibrium price vector* for the initial endowments $\omega \in \overline{\Omega}$ if

$$f(p; \langle\omega_1, p\rangle, \ldots, \langle\omega_m, p\rangle) = \sum_{i=1}^{m} \omega_i. \tag{2.6}$$

It is clear that if p is an equilibrium price vector, then, for $\alpha \in I\!R^1_+$, αp is also an equilibrium price vector.

Using methods similar to those employed in Balasko (1975), Debreu (1970) and Dierker (1974), one can show that, under our assumptions (which are somewhat weaker than in the papers above), for each allocation of initial resources $\omega \in \overline{\Omega}$ there exists an equilibrium price vector. Observe that when $\omega \in (\overline{I\!R}^l_+)^m \setminus \overline{\Omega}$, $\omega \neq 0$ (i.e. some, but not all commodities are not available in the economy), then there is no equilibrium.

2.4 Let $\overline{W} \subset \overline{\Omega} \times I\!R^l_+$ be given by

$$\overline{W} = \{(\omega, p) \mid \omega \in \overline{\Omega},\ p \text{ is an equilibrium price vector at } \omega\},$$

and let $\overline{\pi} : \overline{W} \to \overline{\Omega}$ be the map induced by the projection onto the first factor. Following Debreu, we will call $\overline{W}$ the *Walras correspondence*. The meaning of the word "correspondence" is that over each point $\omega \in \overline{\Omega}$ there lies the set of equilibrium prices at ω. From (2.6) it is clear that $\overline{W}$ is defined in $\overline{\Omega} \times I\!R^l_+$ by the following l equations:

$$\zeta^j(\omega; p) = f^j(p; \langle\omega_1, p\rangle, \ldots, \langle\omega_m, p\rangle) - \sum_{i=1}^{m} \omega_i^j = 0, \quad j = 1, \ldots, l. \tag{2.7}$$

The function $\zeta = (\zeta^1, \ldots, \zeta^l)$ is called the (aggregate) *excess demand function.*

According to (2.2), the equations (2.7) are not independent, but satisfy *Walras law*, i.e., they are subject to the linear relation

$$\langle p, \zeta(\omega, p)\rangle = 0. \tag{2.8}$$

2.5 It is not very convenient to deal with $\overline{\Omega}$ and $\overline{W}$ since they are not manifolds. Of course, as it is usually done, one could limit oneself to considering their interiors, but this would force one to lay aside many important examples. Therefore we proceed in a different way.

First of all, we observe that a participant without initial endowments cannot buy or sell anything and therefore can be excluded from the economy. Thus one can replace the set of admissible initial endowments by its subset

$$\overline{\Omega}^* = \{\omega \in \overline{\Omega} \mid \omega_i \neq \{0\},\ i = 1, \dots, m\}. \tag{2.9}$$

Let $\omega \in \overline{\Omega}^*$. Then it is not hard to see that there exists a neighborhood V_ω of the set $\bar{\pi}^{-1}(\omega)$ in $(\mathbb{R}^l)^m \times \mathbb{R}^l_+$ such that the function $\zeta(\omega; p)$ is defined in V_ω (and satisfies the relation (2.8)). Put

$$V = \bigcup_{\omega \in \overline{\Omega}^*} V_\omega \subset (\mathbb{R}^l)^m \times \mathbb{R}^l_+,$$

and let W be defined in V by the system of l equations (2.7) satisfying the linear relation (2.8). Let $\pi : W \to (\mathbb{R}^l)^m$ be the restriction of the projection $(\mathbb{R}^l)^m \times \mathbb{R}^l_+ \to (\mathbb{R}^l)^m$ on W, and put $\Omega = \pi(W)$. It is clear that

$$\pi^{-1}(\overline{\Omega}^*) = \overline{W}^* = \bar{\pi}^{-1}(\overline{\Omega}^*); \quad \pi|_{\overline{\Omega}^*} = \bar{\pi}|_{\overline{\Omega}^*} \tag{2.10}$$

(here, as usual, $\phi|_X$ denotes the restriction of map ϕ on a subset X).

2.6 Proposition *Let $1 \leq i \leq m$. Then the functions p^j, $w_i = \langle \omega_i, p \rangle$, ω_r^j, $1 \leq j \leq l$; $1 \leq r \leq m$, $r \neq i$ form a system of local (and global) coordinates at each point of W. In particular, W is an $(lm+1)$-dimensional C^1-manifold and π is a C^1-mapping. Moreover, Ω is open in $(\mathbb{R}^l)^m$ and if*

$$W \supset W^n = \{(\omega, p) \in W \mid \sum_{j=1}^{l} p^j = 1\}, \quad \pi^n = \pi|_{W^n}, \tag{2.11}$$

then $\pi^n : W^n \to \Omega$ is a surjective proper map of degree 1.

Proof. Consider the map

$$\sigma : V \to Z = \mathbb{R}^l_+ \times \mathbb{R}^1_+ \times (\mathbb{R}^l)^{m-1} \tag{2.12}$$

defined by the functions from the statement of our proposition. For a point $z \in Z$, $z = (p; w_i; \omega_1, \dots, \omega_{i-1}, \omega_{i+1}, \dots, \omega_m)$ we put

$$\begin{aligned} &\tau(z) = (\omega_1, \dots, \omega_{i-1}, \omega_i(z), \omega_{i+1}, \dots, \omega_m; p), \\ &\tau : Z \to \mathbb{R}^{lm} \times \mathbb{R}^l_+, \end{aligned} \tag{2.13}$$

where

$$\omega_i^j(z) =$$

$$f^j(p; \langle\omega_1, p\rangle, \ldots, \langle\omega_{i-1}, p\rangle, w_i, \langle\omega_{i+1}, p\rangle, \ldots, \langle\omega_m, p\rangle) - \sum_{r \neq i} \omega_r^j. \qquad (2.14)$$

It is clear that

$$\tau \circ \sigma|_W = \mathrm{id}_W : W \to W, \qquad (2.15)$$

where, as usual, $\mathrm{id}_W : W \to W$ denotes the identity map. From (2.14), (2.15) and the definition of W it follows that $\sigma(W)$ is open in Z and that $\sigma|_W : W \to \sigma(W)$ is a C^1-diffeomorphism. This proves the first part of Proposition 2.6.

Next we observe that there is a commutative diagram

$$\begin{array}{ccc} & W & \\ \pi \swarrow & & \searrow \sigma \\ \Omega & \underset{\rho}{\longleftarrow} & \sigma(W) \end{array} \qquad (2.16)$$

where

$$\rho(z) = (\omega_1, \ldots, \omega_{i-1}, \omega_i(z), \omega_{i+1}, \ldots, \omega_m). \qquad (2.17)$$

Since, in view of (2.1), the maps f_i are open and since a composition of open maps is itself open, we see that $\Omega = \rho(\sigma(W))$ is open in $(\mathbb{R}^l)^m$. The remaining claims of Proposition 1.6 are proved in Balasko (1975, §5). □

2.7 We recall that an economy with initial endowments ω is called *regular* if one of the following equivalent conditions holds (cf. Balasko, 1975; Debreu, 1970):

a) In a neighborhood of ω, π^n is an unramified covering (i.e. a proper map which is a diffeomorphism in a neighborhood of each point $z \in (\pi^n)^{-1}(\omega)$; since from 2.6 we already know that π^n is proper, this condition holds iff for each $z \in (\pi^n)^{-1}(\omega)$ the differential $(d\pi^n)_z$ is an isomorphism of tangent spaces);

b) In a neighborhood of ω, the map π is a trivial bundle. In other words, for each $z \in \pi^{-1}(\omega)$ the differential $(d\pi)_z$ is surjective.

In what follows we need explicit formulae for the differential $d\pi$. We fix a number i, $1 \leq i \leq m$ and compute the matrix of the differential $d\pi$ taking p^j, w_i, and ω_r^j as local coordinates in W (cf. 1.6) and ω_i^j, ω_r^j $(j = 1, \dots, l,$ $r = 1, \dots, i-1, i+1, \dots, m)$ as local coordinates in Ω. Differentiating (2.14), we immediately see that

$$d\pi = \begin{pmatrix} S^i & * \\ 0 & \mathrm{id}_{l(m-1)} \end{pmatrix}, \tag{2.18}$$

where $d\pi$ is a matrix of order $lm \times (lm+1)$, $*$ is a matrix of order $l \times l(m-1)$, $\mathrm{id}_{l(m-1)}$ is the identity matrix of order $l \times (m-1)$, and S^i is the $l \times (l+1)$-matrix for which

$$S_{jt}^i = \begin{cases} \frac{\partial f^j}{\partial p^t} + \sum_{r \neq i} \frac{\partial f_r^j}{\partial w_r} \cdot \omega_r^t = \frac{\partial \zeta^j}{\partial p^t} - \frac{\partial f_i^j}{\partial w_i} \cdot \omega_i^t, \ 1 \leq j, t \leq l; \\ \frac{\partial f_i^j}{\partial w_i}, \ 1 \leq j \leq l, \ t = l+1. \end{cases} \tag{2.19}$$

2.8 Proposition *The following conditions are equivalent:*

a) $\omega \in \Omega$ *is a regular economy;*

b) $\mathrm{rk}(S^i)^{l \times (l+1)} = l$, *i.e. the rows of* S^i *are linearly independent;*

c) $\mathrm{rk}((\frac{\partial \zeta}{\partial p})^{l \times l} - (\frac{\partial f_i}{\partial w_i})^{l \times 1} \cdot \omega_i^{1 \times l}) = l$;

d) $\mathrm{rk}(\frac{\partial \zeta}{\partial p})^{l \times l} = l - 1$.

Proof. a) $\Leftrightarrow$ b) immediately follows from (2.18).
b) $\Leftrightarrow$ c). We observe that by (2.19)

$$S^i = \left(\left(\frac{\partial \zeta}{\partial p} - \frac{\partial f_i}{\partial w_i} \cdot \omega_i \right)^{l \times l} ; \frac{\partial f_i}{\partial w_i} \right)^{l \times (l+1)}. \tag{2.20}$$

Since ζ is a homogeneous function of degree zero of p (cf. (2.3)), the Euler formula yields

$$\frac{\partial \zeta}{\partial p} \cdot p = 0. \tag{2.21}$$

From (2.20) and (2.21) it follows that

$$S^i = \left(\frac{\partial \zeta}{\partial p} - \frac{\partial f_i}{\partial w_i} \cdot \omega_i \right)^{l \times l} \cdot \left(\mathrm{id}_l ; -\frac{p}{w_i} \right)^{l \times (l+1)}, \tag{2.22}$$

where id_l is the identity $l \times l$-matrix and $w_i = \langle \omega_i, p \rangle$.
Relation (2.22) shows that

$$\mathrm{rk} S^i = \mathrm{rk}\left(\frac{\partial \zeta}{\partial p} - \frac{\partial f_i}{\partial w_i} \cdot \omega_i \right), \tag{2.23}$$

which proves that conditions b) and c) are equivalent.

b) $\Leftrightarrow$ d). From (2.20) it follows that

$$\mathrm{rk} S^i = \mathrm{rk}\left(\frac{\partial \zeta}{\partial p}; \frac{\partial f_i}{\partial w_i} \right). \tag{2.24}$$

Differentiating (2.8) with respect to p, we get

$$p \cdot \frac{\partial \zeta}{\partial p} + \zeta = 0. \tag{2.25}$$

Since p is an equilibrium price vector for the initial endowments ω, one has

$$p \cdot \frac{\partial \zeta}{\partial p} = 0. \tag{2.26}$$

On the other hand, differentiating (2.2) with respect to w_i, we get the so called Engel aggregation condition (cf. Intriligator, 1971, (7.4.38)):

$$\left\langle p, \frac{\partial f_i}{\partial w_i} \right\rangle = 1. \tag{2.27}$$

From (2.26) and (2.27) it is clear that the vector $\frac{\partial f_i}{\partial w_i}$ does not lie in the linear span of the vectors $\frac{\partial \zeta}{\partial p^j}$, $j = 1, \ldots, l$. Therefore

$$\mathrm{rk} S^i = \mathrm{rk}\left(\frac{\partial \zeta}{\partial p} \right) + 1, \tag{2.28}$$

which proves the equivalence of conditions b) and d). This concludes the proof. □

2.9 Let $\omega \in \overline{\Omega}^*$, and let $1 \leq i \leq m$. Put

$$\begin{aligned} \Omega_i &= \{\omega' \in \Omega \mid \omega'_r = \omega_r,\ r \neq i\}, \\ \overline{\Omega}_i &= \{\omega' \in \overline{\Omega} \mid \omega'_r = \omega_r,\ r \neq i\}. \end{aligned} \tag{2.29}$$

Without loss of generality we may assume that Ω_i is a connected manifold of dimension l. Put $W_i = \pi^{-1}(\Omega_i)$, and let $\pi_i = \pi|_{W_i} : W_i \to \Omega_i$. $\overline{W}_i$ and $\bar{\pi}_i$ are defined in a similar way. From 2.6 it is easy to deduce that W_i is a connected C^1-manifold of dimension $l+1$ and π_i is a surjective map. It is clear that

$$d\pi_i = S^i, \tag{2.30}$$

where $p^j; w_i$ are chosen as local coordinates in W_i, ω_i^j $(j = 1, \dots, l)$ are chosen as local coordinates in Ω_i, and S^i is the $l \times (l+1)$-matrix defined in (2.19).

From this and 2.8 it follows that $\omega \in \Omega_i \subset \Omega$ is a regular value of π_i iff ω is a regular value of π, i.e. the corresponding economy is regular.

We observe that if ω is a regular economy, then, in some neighborhood $U_i \ni \omega$, W_i splits into several nonintersecting branches:

$$\pi_i^{-1}(U_i) = V_{i1} \cup \ldots \cup V_{is}; \quad V_{ir} \cap V_{it} = \emptyset, \ r \neq t, \tag{2.31}$$

where

$$V_{ik} = V_{ik}^n \cdot I\!R_+^1, \quad k = 1, \dots, s \tag{2.32}$$

(cf. (2.11)) and the map

$$\pi_i|_{V_{ik}^n} : V_{ik}^n \to U_i, \quad k = 1, \dots, s \tag{2.33}$$

is a C^1-diffeomorphism.

2.10 Remark The assumption that Ω is obtained by extending $\overline{\Omega}^*$ is not essential, and the above results easily generalize to more general domains. However economists usually consider orthants (or Edgeworth boxes), and the conditions for infinitesimal stability on the boundary assume a nicer form in this case (cf. §2).

3 Infinitesimal stability

3.1 Let $\omega \in \overline{\Omega}^*$ be a regular economy. Let p be an equilibrium price vector in ω, and let

$$(\omega, p) \in V_{it}, \tag{3.1}$$

where $1 \leq t = t(p) \leq s$ (cf. (2.31)). For an arbitrary $\omega' \in U_i$ we define the indirect utility of the i-th participant as

$$v_i^p(\omega') = u_i(f_i(p(\omega'), \langle \omega_i', p(\omega') \rangle)), \tag{3.2}$$

where $p(\omega')$ is the only (modulo a positive factor) equilibrium price vector for the initial endowments ω' such that

$$(\omega', p(\omega')) \in V_{it} \tag{3.3}$$

(here we used the fact that f_i is homogeneous of degree zero). We observe that $p(\omega')$ and hence $v_i^p(\omega')$ is a function of class C^1.

3.2 Definition

(a) The equilibrium distribution corresponding to regular initial endowments $\omega \in \overline{\Omega}^*$ and equilibrium price vector p is called *infinitesimally stable* (resp. *weakly infinitesimally stable*) with respect to the i-th participant if

$$\frac{\partial v_i^p}{\partial \omega_i^j}(\omega) > 0 \tag{3.4}$$

(resp.

$$\frac{\partial v_i^p}{\partial \omega_i^j}(\omega) \geq 0) \tag{3.5}$$

for all j for which $\omega_i^j > 0$. In case (3.4) we write (ω, p) i.s. (i) and in case (3.5) (ω, p) w.i.s. (i).

(b) The equilibrium distribution corresponding to regular initial endowments $\omega \in \overline{\Omega}^*$ and an equilibrium price vector p is called *absolutely infinitesimally unstable* (resp. *strongly absolutely infinitesimally unstable*) with respect to the i-th participant if

$$\frac{\partial v_i^p}{\partial \omega_i^j}(\omega) \leq 0 \tag{3.6}$$

(resp.

$$\frac{\partial v_i^p}{\partial \omega_i^j}(\omega) < 0) \tag{3.7}$$

for all j for which $\omega_i^j > 0$. In case (3.6) we write (ω, p) a.i.us. (i), in case (3.7) (ω, p) s.a.i.us. (i).

In the case when $s = 1$ in (2.31) we say that ω has one of the properties defined in (a) or (b) above if the only equilibrium distribution corresponding to ω has this property.

We notice that in order for Definition 3.2 to make sense it suffices that the differential $d\pi_i$ be surjective at the point $(p, \langle \omega_i, p \rangle)$. The results below hold also in this more general case, but for our purposes it suffices to consider regular economies (a less trivial generalization of Definition 3.2 will be given in 3.15).

3.3 Remark Economically speaking, condition (3.4) (resp. condition (3.7)) implies that the indirect utility of participant i at equilibrium price p is increasing (resp. decreasing) in the initial endowments of i. We observe that the properties of infinitesimal stability and strong absolute infinitesimal instability are open. More precisely, let $\omega \in \overline{\Omega}^*$, $\omega_i > 0$ be a regular economy such that (ω, p) i.s. (i) (resp. (ω, p) s.a.i.us. (i)) for each equilibrium price vector p at ω, and let $\omega' \in \overline{\Omega}^*$ be an allocation sufficiently close to ω. Then (ω', p') i.s. (i) (resp. (ω', p') s.a.i.us. (i)) for each equilibrium price vector p' at ω'.

3.4 To compute the gradient of the function v_i^p it is convenient to pass to the local coordinates introduced in 2.9. We have:

$$v_i(p, w_i) \stackrel{\text{def}}{=} u_i(f_i(p, w_i)), \tag{3.8}$$

$$\frac{\partial v_i}{\partial p}(p, w_i) = u_i'(f_i(p, w_i)) \cdot \frac{\partial f_i}{\partial p}(p, w_i), \tag{3.9}$$

$$\frac{\partial v_i}{\partial w_i}(p, w_i) = u_i'(f_i(p, w_i)) \cdot \frac{\partial f_i}{\partial w_i}(p, w_i), \tag{3.10}$$

and from (2.1) it follows that

$$u_i'(f_i(p, w_i)) = \text{grad } u_i(f_i(p, w_i)) = \lambda p, \tag{3.11}$$

where $\lambda = \lambda(p, w_i)$ is a positive scalar.

Differentiating by p the Walras relation (2.2), we get

$$p^{1\times l} \cdot \left(\frac{\partial f_i}{\partial p}\right)^{l\times l} + (f_i)^{1\times l} = 0 \tag{3.12}$$

(this is the so called *Cournot aggregation condition*; cf. Intriligator, 1971, (7.4.41)).

Combining (3.11), (3.12), and (2.27), we see that

$$dv_i = -\lambda \sum_{j=1}^{l} f_i^j dp^j + \lambda dw_i. \tag{3.13}$$

Let

$$\vartheta = \sum_{j=1}^{l} -\alpha_j \frac{\partial}{\partial p^j} + \beta \frac{\partial}{\partial w_i}. \tag{3.14}$$

From (3.13) it follows that

$$\frac{\partial v_i}{\partial \vartheta} > 0 \text{ (resp. } \geq 0) \;\Leftrightarrow\; \langle f_i, \alpha \rangle + \beta > 0 \text{ (resp. } \geq 0). \tag{3.15}$$

3.5 We recall the definition of normal demand.

Definition A demand function $f_i(p, w_i)$ is called *normal* (resp. *strictly normal*) if

$$\frac{\partial f_i}{\partial w_i} \geq 0 \tag{3.16}$$

(resp.

$$\frac{\partial f_i}{\partial w_i} > 0) \tag{3.17}$$

for all $p \in I\!R^l_+$, $w_i \in \overline{I\!R}^1_+$.

Normality of demand means the absence of commodities that have little value for the i-th participant (cf. Intriligator, 1971, (7.4.29)); for example it excludes Giffen commodities.

3.6 Theorem *Suppose that the i-th economic agent has normal demand and positive initial endowments* (*i.e.* $\omega_i > 0$). *Then there are no equilibrium allocations that are absolutely infinitesimally unstable with respect to the i-th participant.*

Proof. According to (2.19),

$$d\pi \left(\frac{\partial}{\partial w_i} \right) = \sum_{j=1}^{l} \frac{\partial f_i^j}{\partial w_i} \frac{\partial}{\partial \omega_i^j}. \tag{3.18}$$

On the other hand, from (3.15) it follows that

$$\frac{\partial v_i}{\partial w_i} = \sum_{j=1}^{l} \frac{\partial f_i^j}{\partial w_i} \cdot \frac{\partial v_i^p}{\partial \omega_i^j} > 0. \tag{3.19}$$

Since the demand is normal (cf. (3.16)), there exists a commodity $1 \leq j \leq l$ for which

$$\frac{\partial v_i^p}{\partial \omega_i^j} > 0, \tag{3.20}$$

which contradicts (3.6) (because $\omega_i > 0$). This proves the theorem. □

3.7 Next we prove a criterion for infinitesimal stability.

Theorem *Let $\omega \in \overline{\Omega}^*$ be a regular economy, and let p be an equilibrium price vector at ω. Then (ω, p) is i.s. (i) iff the system of equations*

$$(V^i(\omega, p))^{l\times l} \cdot x^{l\times 1} = (f_i(p, w_i))^{l\times 1}, \tag{3.21}$$

where

$$\begin{aligned}(V^i_{j,t}(\omega, p)) &= -\frac{\partial \zeta^t}{\partial p^j}(\omega, p) + \frac{\partial f_i^t}{\partial w_i}(p, \langle \omega_i, p\rangle) \cdot \omega_i^j \\ &= -\frac{\partial f^t}{\partial p^j}(p; \langle \omega_1, p\rangle, \dots, \langle \omega_m, p\rangle) \\ &\quad -\sum_{r\neq i} \frac{\partial f_r^t}{\partial w_r}(p; \langle \omega_r, p\rangle) \cdot \omega_r^j, \quad j, t = 1, \dots, l,\end{aligned} \tag{3.22}$$

has a solution $x = (x^1, \dots, x^l)$ such that

$$x^j > 0 \text{ for all } j \text{ for which } \omega_i^j > 0. \tag{3.23}$$

Proof. Let

$$\frac{\partial}{\partial \omega_i^j}(\omega) = \sum_{r=1}^{l} -\alpha_{rj}\frac{\partial}{\partial p^r} + \beta_j \frac{\partial}{\partial w_i}, \quad j = 1, \dots, l. \tag{3.24}$$

In view of (3.15),

$$\frac{\partial v_i^p}{\partial \omega_i^j}(\omega) > 0 \Leftrightarrow \langle f_i(p, w_i), \alpha_j\rangle + \beta_j > 0. \tag{3.25}$$

Taking into consideration (2.19), (2.30), and (3.25), we see that $\frac{\partial v_i^p}{\partial \omega_i^j}(\omega) > 0$ iff the system of equations and inequalities

$$\begin{cases} (V^i(\omega, p))' \cdot (\alpha_j)^{l\times 1} + \beta_j \left(\frac{\partial f_i}{\partial w_i}(p, \omega_i)\right)^{l\times 1} = (\delta_j)^{l\times 1}; \\ \langle f_i(p, w_i), \alpha_j\rangle + \beta_j > 0, \end{cases} \tag{3.26}$$

where $(V^i(\omega,p))'$ is the transpose of $V^i(\omega,p)$ and $(\delta_j)^k = \delta_{jk}$, $k = 1,\ldots,l$ (here δ_{jk} is the Kronecker delta), has a solution.

From (3.22) and (2.21) it follows that

$$(V^i(\omega,p))' \cdot p = w_i \cdot \frac{\partial f_i}{\partial w_i}(p,w_i). \tag{3.27}$$

Put

$$\gamma_j = \alpha_j + \frac{\beta_j}{w_i} \cdot p. \tag{3.28}$$

Combining (3.26), (3.28) and (2.2), we see that

$$\begin{cases} (V^i(\omega,p))' \cdot \gamma_j = (V^i(\omega,p))' \cdot \alpha_j + \frac{\beta_j}{w_i} \cdot (V^i(\omega,p))' \cdot p = \delta_j; \\ \langle f_i(p,w_i), \gamma_j\rangle = \langle f_i(p,w_i), \alpha_j\rangle + \frac{\beta_j}{w_i}\langle f_i(p,w_i), p\rangle \\ \qquad = \langle f_i(p,\omega_i), \alpha_j\rangle + \beta_j > 0. \end{cases} \tag{3.29}$$

Thus the system (3.26) has a solution α_j iff the system (3.29) has a solution γ_j (this also immediately follows from the fact that one can introduce a norm for which $w_i \equiv 1$).

Since ω is regular, Proposition 2.8 shows that the matrix $(V^i(\omega,p))'$ is invertible, so that the system (3.29) has a unique solution

$$\gamma_j = ((V^i(\omega,p))')^{-1}\delta_j, \quad j = 1,\ldots,l. \tag{3.30}$$

Thus, for infinitesimal stability it is necessary and sufficient that for all j for which $\omega_i^j > 0$ there is an inequality

$$[f_i'(p,w_i)]^{1\times l} \cdot [((V^i(\omega,p))')^{-1} \cdot \delta_j]^{l\times 1} > 0. \tag{3.31}$$

But $f_i'(p,w_i) \cdot ((V^i(\omega,p))')^{-1}$ is the j-th coordinate of the row vector

$$x' = \left(f_i'(p,w_i)^{1\times l} \cdot ((V^i(\omega,p))'\right)^{-1} \tag{3.32}$$

such that x is the unique (since V^i is invertible) solution of the system of equations (3.21). This proves the theorem. □

3.8 An almost word for word repetition of the proof of Theorem 3.7 yields the following result.

Theorem *Let $\omega \in \overline{\Omega}^*$ be a regular economy and let p be an equilibrium price vector at ω.*

a) *(ω, p) is w.i.s. (i) iff the system of equations* (3.21) *has a solution x such that*

$$x^j \geq 0 \text{ for all } j \text{ for which } \omega_i^j > 0. \tag{3.33}$$

b) *(ω, p) is a.i.us. (i) (resp. s.a.i.us. (i)) iff the system of equations* (3.21) *has a solution x such that*

$$x^j \leq 0 \text{ for all } j \text{ for which } \omega_i^j > 0. \tag{3.34}$$

(resp.

$$x^j < 0 \text{ for all } j \text{ for which } \omega_i^j > 0). \tag{3.35}$$

3.9 For some applications it is more convenient to restate the criteria given in Theorems 3.7 and 3.8 in terms of excess demand functions.

Let

$$\zeta_i(\omega, p) = f_i(p, \langle \omega_i, p \rangle) - \omega_i \tag{3.36}$$

be the excess demand function of the i-th participant.

Theorem *Let $\omega \in \overline{\Omega}^*$ be a regular economy, and let p be an equilibrium price vector at ω. Then (ω, p) is i.s. (i) (resp. (ω, p) is w.i.s. (i), resp. (ω, p) is a.i.us. (i), resp. (ω, p) is s.a.i.us. (i)) iff the system of equations*

$$\begin{cases} x^{1\times l} \left(\frac{\partial \zeta}{\partial p}(\omega, p) \right)^{l \times l} + (\zeta_i(\omega, p))^{1 \times l} = 0; \\ \left\langle x, \frac{\partial f_i}{\partial w_i}(p, \langle \omega_i, p \rangle) \right\rangle = 1 \end{cases} \tag{3.37}$$

has a solution satisfying (3.23) *(resp.* (3.33)*, resp.* (3.34)*, resp.* (3.35)*).*

We observe that, in view of (2.21), the first l equations of (3.37) are linearly dependent, and thus one may assume that (3.37) is a system of l equations with l unknowns (from 2.8 it follows that the rank of this system of equations is equal to l).

Proof. In view of Theorems 3.7 and 3.8, it suffices to show that the systems (3.37) and (3.21) are equivalent to each other. We have

$$V^i(\omega, p) = \left(\left(-\frac{\partial \zeta}{\partial p}(\omega, p) \right)^{l \times l} + \left(\frac{\partial f_i}{\partial w_i}(p, \langle \omega_i, p \rangle) \right)^{l \times 1} \omega_i^{1 \times l} \right)'. \tag{3.38}$$

Suppose first that x satisfies (3.37). Then

$$\begin{aligned} V^i x' &= -\left(\frac{\partial \zeta}{\partial p}\right)' x' + \left\langle \frac{\partial f_i}{\partial w_i}, x \right\rangle \omega_i \\ &= -\zeta_i(\omega, p) + \omega_i = (f_i(p, \langle \omega_i, p \rangle))^{l\times 1}. \end{aligned} \tag{3.39}$$

Conversely, if x satisfies (3.21), then

$$\left(\frac{\partial \zeta}{\partial p}\right)' x + f_i(p, \langle \omega_i, p \rangle) - \left\langle x, \frac{\partial f_i}{\partial w_i}(p, \langle \omega_i, p \rangle) \right\rangle \omega_i = 0. \tag{3.40}$$

Since, in view of (2.21),

$$p \perp \mathrm{Im}\left(\frac{\partial \zeta}{\partial p}(\omega, p)\right)', \tag{3.41}$$

i.e. the image of the linear operator $\frac{\partial \zeta}{\partial p}(\omega, p)'$ lies in the orthogonal complement to the vector p, one has

$$\langle \omega_i, p \rangle \cdot \left\langle x, \frac{\partial f_i}{\partial w_i}(p, \langle \omega_i, p \rangle) \right\rangle - \langle f_i(p, \langle \omega_i, p \rangle), p \rangle = 0, \tag{3.42}$$

i.e.

$$w_i \left\langle x, \frac{\partial f_i}{\partial w_i}(p, \langle \omega_i, p \rangle) \right\rangle = w_i, \quad \left\langle x, \frac{\partial f_i}{\partial w_i}(p, \langle \omega_i, p \rangle) \right\rangle = 1. \tag{3.43}$$

Since (3.37) is obtained by combining (3.40) and (3.43), this completes the proof. □

3.10 Theorem *Suppose that a regular economy $\omega \in \overline{\Omega}^*$ is an equilibrium allocation for the i-th participant for an equilibrium price vector p, i.e.*

$$\zeta_i(\omega, p) = \zeta(\omega, p) = 0. \tag{3.44}$$

Then (ω, p) is i.s. (i).

Proof. In view of (2.26) and (2.27), to prove the theorem it is sufficient to put $x = p$ in Theorem 3.9 (we recall that under our assumptions $p > 0$). □

Theorem 3.10 says that the equilibrium is i.s. (i) provided that at the equilibrium participant i does not trade, i.e. his consumption equals his initial endowment. One can be more precise if this condition holds for all participants.

3.11 Corollary *Suppose that an allocation ω is Pareto optimal. Then ω is infinitesimally stable with respect to each of the participants (cf. also 3.12).*

3.12 Remark It is well known (cf. e.g. Balasko, 1975) that any Pareto optimal allocation ω is regular and admits a unique (modulo a positive factor) equilibrium price vector. Furthermore, from 2.1 it follows that $\omega_i > 0, i = 1, \ldots, m$. Therefore from Remark 3.3 and Corollary 3.11 it follows that all allocations sufficiently close to ω are infinitesimally stable with respect to each of the participants.

3.13 Corollary *Suppose that all economic agents are sufficiently close to each other in the sense that $\omega_i - \omega_k$ are small vectors and $f_i - f_k$ are small functions (with respect to a norm in the space of continuously differentiable functions on suitable compacts) for all $1 \leq i, k \leq m$. Then ω is a regular economy, there exists a unique (modulo a positive factor) equilibrium price vector at ω, and ω is infinitesimally stable with respect to all the participants.*

Thus infinitesimally unstable economies arise only if the economic inequality between the agents is sufficiently high.

3.14 Since in the case when $\omega \in \partial\overline{\Omega}^*$ not all the partial derivatives are involved in Definition 3.2, it is natural to slightly generalize Definition 3.2 and Theorems 3.7–3.9 by considering some economies that fail to be regular.

Definition Let $\omega \in \overline{\Omega}^*$, and let p be an equilibrium price vector at ω. The couple (ω, p) will be called *regular* with respect to the i-th participant and j-th commodity ((ω, p) reg. (i/j)) if

$$\frac{\partial}{\partial \omega_i^j} = d\pi_i(\vartheta), \tag{3.45}$$

where $\frac{\partial}{\partial \omega_i^j}$ is a vector from the standard basis of the tangent space to the linear space $I\!R^{lm}$ and ϑ is a tangent vector to W_i at the point $(p, \langle \omega_i, p \rangle)$.

We observe that if ω is a regular economy, then for each equilibrium price vector p at ω, each participant $1 \leq i \leq m$, and each commodity $1 \leq j \leq l$ the couple (ω, p) is reg. (i/j). Conversely, from 2.9 it follows that if for each equilibrium price vector p at ω there exists a participant $1 \leq i \leq m$ such that (ω, p) is reg. $i/(1, \ldots, l)$, then ω is a regular economy.

3.15 Definition Let (ω, p) be reg. (i/j). The equilibrium corresponding to ω and p is called *infinitesimally stable* (resp. *infinitesimally unstable*, resp. *strongly infinitesimally unstable*) with respect to the i-th participant and j-th commodity if for each ϑ satisfying (3.45) one has

$$\frac{\partial v_i^p}{\partial \vartheta}(p, \langle \omega_i, p \rangle) > 0 \tag{3.46}$$

(resp.

$$\frac{\partial v_i^p}{\partial \vartheta}(p, \langle \omega_i, p \rangle) \geq 0, \tag{3.47}$$

resp.

$$\frac{\partial v_i^p}{\partial \vartheta}(p, \langle \omega_i, p \rangle) \leq 0, \tag{3.48}$$

resp.

$$\frac{\partial v_i^p}{\partial \vartheta}(p, \langle \omega_i, p \rangle) < 0). \tag{3.49}$$

In the case (3.46) (resp. (3.47), resp. (3.48), resp. (3.49)) we will use abbreviation (ω, p) i.s. (i/j) (resp. (ω, p) w.i.s. (i/j), resp. (ω, p) i.us. (i/j), resp. (ω, p) s.i.us. (i/j)).

We observe that if $\omega \in \overline{\Omega}^*$ is a regular economy, then (ω, p) is i.s. $i \Leftrightarrow$ (ω, p) is i.s. $(i/1, \ldots, l)$; (ω, p) is w.i.s. $(i) \Leftrightarrow (\omega, p)$ is w.i.s. $(i/1, \ldots, l)$; (ω, p) is a.i.us. $(i) \Leftrightarrow (\omega, p)$ is i.us. $(i/1, \ldots, l)$; (ω, p) is s.a.i.us. $(i) \Leftrightarrow (\omega, p)$ is s.i.us. $(i/1, \ldots, l)$.

For the sake of brevity, in what follows we will consider only *infinitesimal stability* with respect to the i-th participant and j-th commodity. However the reader will easily restate Theorems 3.16–3.18 for the other notions introduced in 3.15 (the corresponding theorems are in the same relation to Theorems 3.16–3.18 as Theorem 3.8 is to Theorem 3.7).

3.16 Theorem *Suppose that (ω, p) is reg. (i/j). Then the following conditions are equivalent:*

a) *(ω, p) is i.s. (i/j);*

b) *The system of equations* (3.21) *has a solution x for which $x^j > 0$;*

c) *$x^j > 0$ for each solution x of the system* (3.21).

Proof. Arguing as in the proof of Theorem 3.7, we see that (ω, p) is i.s. (i/j) iff each solution of the system of linear equations

$$(V^i(\omega, p))' \cdot \gamma_j = \delta_j \tag{3.50}$$

satisfies the inequality

$$\langle f_i(p, w_i), \gamma_j \rangle > 0 \tag{3.51}$$

(cf. (3.15), (3.29)).

a) $\Rightarrow$ b). We observe that for each solution γ_j of the system (3.50) (solutions exist in view of (3.45)) and each $\xi \in \mathrm{Ker}(V^i(\omega,p))'$ the vector $\gamma_j + \xi$ is also a solution of (3.50) (here and in what follows Ker$A$ denotes the kernel of linear operator A, i.e. the linear subspace of vectors on which A vanishes). From this it follows that if the vector $f_i(p,w_i)$ is not orthogonal to $\mathrm{Ker}(V^i(\omega,p))'$, then (3.50) has a solution that does not satisfy (3.51). Thus from a) it follows that

$$f_i(p,w_i) \perp \mathrm{Ker}(V^i(\omega,p))'. \tag{3.52}$$

It is well known (and easy to show) that (3.52) is equivalent to the condition

$$f_i(p,w_i) \in \mathrm{Im}V^i(\omega,p), \tag{3.53}$$

so that a) implies that the system of equations (3.21) has a solution.

Let

$$V^i(\omega,p)\cdot x = f_i(p,w_i), \quad x \in I\!\!R^l. \tag{3.54}$$

Then from (3.51) it follows that

$$\begin{aligned} 0 \; < \; & \langle f_i(p,w_i),\gamma_j\rangle = \langle V^i(\omega,p)\cdot x,\gamma_j\rangle \\ & = \langle x,(V^i(\omega,p))'\cdot\gamma_j\rangle = \langle x,\delta_j\rangle = x^j. \end{aligned} \tag{3.55}$$

The implication c) $\Rightarrow$ b) is obvious.

To prove the implication b) $\Rightarrow$ a) it suffices to read the chain of equalities (3.55) from the right to the left. This concludes the proof. □

3.17 Theorem *Suppose that (ω,p) is reg. (i/j). Then the following conditions are equivalent:*

a) (ω,p) is i.s. (i/j);

b) The system of equations (3.37) *has a solution x for which $x^j > 0$;*

c) $x^j > 0$ for each solution x of the system (3.37).

Proof. Theorem 3.17 is deduced from Theorem 3.16 in the same way as Theorem 3.9 was deduced from Theorem 3.7. □

3.18 Definitions 3.14 and 3.15 allow to generalize the notion of infinitesimal stability to the case of not necessarily regular boundary economies.

Definition Let $\omega \in \overline{\Omega}^*$, $1 \leq i \leq m$, and put

$$J = \{j \mid \omega_i^j > 0\}. \tag{3.56}$$

Let p be an equilibrium price vector at ω. Suppose that (ω, p) is reg. (i/J), i.e. (ω, p) is reg. (i/j) for all $j \in J$. We say that the equilibrium corresponding to ω and p is *infinitesimally stable* with respect to the i-th participant ((ω, p) is i.s. (i)) if (ω, p) is i.s. (i/J), i.e. (ω, p) is i.s. (i/j) for all $j \in J$.

From Theorems 3.16 and 3.17 it follows that Theorems 3.7 and 3.9 hold also for this more general definition of infinitesimal stability.

3.19 We turn to a study of infinitesimal stability in systems with equilibrium gross substitutability.

Definition An economy ω is said to have the property of (weak) *equilibrium gross substitutability* if

$$\frac{\partial f^j}{\partial p^k}(p; \langle \omega_1, p \rangle, \ldots, \langle \omega_m, p \rangle) \geq 0, \quad j \neq k \tag{3.57}$$

for each *equilibrium* price vector p at ω.

We observe that Definition 3.19 seems less restrictive than the usual definition of the (global weak) gross substitutability in which it is required that the inequalities (3.57) hold for *all* p. On the other hand, usually gross substitutability is considered for *excess* demand functions:

$$\frac{\partial \zeta^j}{\partial p^k}(\omega, p) \geq 0, \quad j \neq k, \tag{3.58}$$

which, at least in the case when all the participants have normal demand, yields weaker restrictions on the preference relations.

3.20 Theorem *Let $\omega \in \overline{\Omega}^*$ be an economy with equilibrium gross substitutability, and let $1 \leq i \leq m$. Suppose that all participants except, possibly, the i-th one, have normal demand (cf. 3.5) and that one of the following conditions holds:*

a) $\omega_i > 0$*;*

b) The i-th participant has strictly normal demand;

c) For each equilibrium price vector p at ω the matrix $(p; \langle \omega_1, p \rangle, \ldots, \langle \omega_m, p \rangle)$ is indecomposable.

(We remark that it suffices to require the demand to be normal or strictly normal only for the equilibrium prices.) Then

A) ω is a regular economy;

B) There exists a unique (up to a constant factor) equilibrium price vector p at ω;

C) ω (or, which is the same, the equilibrium allocation corresponding to (ω, p)) is infinitesimally stable with respect to the i-th participant.

Proof. From our assumptions it follows that

$$v_{r,s} \leq 0, \quad r \neq s \tag{3.59}$$

(cf. (3.22)).

In case a), from (2.27) it follows that

$$V^i \cdot p = \omega_i > 0. \tag{3.60}$$

In case b), (3.27) yields

$$p \cdot V^i = w_i \cdot \frac{\partial f_i}{\partial w_i} > 0. \tag{3.61}$$

In both cases V^i is a matrix with dominant diagonal (cf. Nikaido, 1979, (21.1)), and in particular

$$\det V^i \quad > 0, \tag{3.62}$$

$$(V^i)^{-1} \quad \geq 0. \tag{3.63}$$

(cf. Intriligator, 1971, §6.2).

We claim that the conditions (3.62) and (3.63) are also satisfied in case c). In fact, from our assumptions it follows that the matrix V^i is also indecomposable. From the Perron-Frobenius theorem it follows that if V^i is not invertible, then there exists a vector $x > 0$ such that

$$x \cdot V^i = 0. \tag{3.64}$$

But since $\omega_i \neq 0$, (3.64) contradicts (3.60). Thus V^i is invertible and the properties (3.62) and (3.63) follow from the Perron-Frobenius theorem.

Since V^i is invertible, assertion A) follows from 2.8. Assertion B) follows from (3.62) in view of Proposition 2.6, and assertion C) follows from (3.63) in view of Theorem 3.7. This concludes the proof. □

We observe that if all the participants have normal demand, then from the Perron-Frobenius theorem it immediately follows that the nonzero eigenvalues of the matrix $\left(\frac{\partial \zeta^j}{\partial p^k}\right)$ have negative real parts. Hence the equilibrium corresponding to (ω, p) is locally stable with respect to tâtonnement.

3.21 Our immediate goal consists in strengthening Theorem 3.20 extending it to boundary economies. First we prove the following result.

Theorem. *Let $\omega \in \overline{\Omega}^*$ be an economy with equilibrium gross substitutability, and let J be the subset of the set $\{1, \ldots, l\}$ defined in* (3.56)*. Suppose further that all the participants have normal demand. Then, for each $1 \leq i \leq m$ and each equilibrium price vector p at ω, (ω, p) is reg. (i/J).*

Proof. In view of (3.29) and 3.14, it suffices to show that for $j \in J$

$$\delta_j \in \mathrm{Im}(V^i)'. \tag{3.65}$$

As we already observed in 3.16, (3.65) is equivalent to the condition

$$\delta_j \perp \mathrm{Ker} V^i. \tag{3.66}$$

In other words, it suffices to show that for each $x \in \mathrm{Ker} V^i$ one has

$$x^j = 0, \quad j \in J. \tag{3.67}$$

By the Perron-Frobenius theorem, there exists a nonnegative vector $\gamma_1 \in \mathrm{Ker} V^i$. We put

$$K_1 = \{k \mid \gamma_1^k > 0\}, \quad 1 \leq k \leq l. \tag{3.68}$$

We observe that the matrix V^i is decomposable. More precisely,

$$V^i_{r,s} = 0, \quad r \notin K_1, \ s \in K_1. \tag{3.69}$$

In fact, for $r \notin K_1$ one has

$$(V^i \cdot \gamma_1)^r = \sum_{s=1}^{l} V^i_{r,s} \cdot \gamma_1^s = 0. \tag{3.70}$$

From (3.70) it follows that

$$V^i_{r,r} \cdot \gamma_1^r = \sum_{s \neq r} (-V^i_{r,s}) \cdot \gamma_1^s. \tag{3.71}$$

Since $r \notin K_1$, one has $\gamma_1^r = 0$. Hence from (3.59) it follows that if $\gamma_1^s > 0$, then $V_{r,s}^i = 0$, which yields (3.69). Re-indexing, if necessary, the commodities, one can represent V^i in the form

$$V^i = \begin{pmatrix} A_1 & B_1 \\ 0 & D_1 \end{pmatrix}, \tag{3.72}$$

where the square matrix A_1 corresponds to the commodities from K_1, the square matrix D_1 corresponds to the commodities from $\hat{K}_1$, and

$$K_1 \cap \hat{K}_1 = \emptyset, \quad K \cup \hat{K}_1 = \{1, \dots, l\}. \tag{3.73}$$

From (3.27) it follows that the semi-positive vector $\frac{\partial f_i}{\partial w_i}$ lies in $\mathrm{Im}(V^i)' = (\mathrm{Ker} V^i)^\perp$. Hence

$$\frac{\partial f_i^s}{\partial w_i} = 0, \quad s \in K_1. \tag{3.74}$$

From (3.27) and (3.74) we conclude that (in obvious notations)

$$p^{K_1} \cdot A_1 = 0. \tag{3.75}$$

On the other hand, since the matrix B_1 is non-positive, one has

$$A_1 \cdot p^{K_1} = \omega_i^{K_1} - B_1 \cdot p^{\hat{K}_1} \geq \omega_i^{K_1}. \tag{3.76}$$

From (3.75) and (3.76) it follows that

$$0 = (p^{K_1} \cdot A_1) \cdot p^{K_1} = p^{K_1} \cdot (A_1 \cdot p^{K_1}) \geq \langle p^{K_1}, \omega_i^{K_1} \rangle. \tag{3.77}$$

Since $p^{K_1} > 0$, (3.77) implies that

$$\omega_i^{K_1} = B_1 \cdot p^{\hat{K}_1} = 0, \tag{3.78}$$

i.e.

$$J \cap K_1 = \emptyset.$$

Next we observe that D_1 is a matrix of the same type as V^i. Suppose that

$$D_1 \cdot \gamma_2 = 0, \quad \gamma_2 \geq 0, \tag{3.79}$$

and put

$$K_2 = \{k \in \hat{K}_1 \mid \gamma_2^k > 0\}. \tag{3.80}$$

Then, after a suitable renumbering of commodities, D_1 can be represented in the form

$$D_1 = \begin{pmatrix} A_2 & B_2 \\ 0 & D_2 \end{pmatrix}, \tag{3.81}$$

where the matrix A_2 corresponds to the commodities from K_2.

From (3.78) it follows that

$$B_1 = 0, \tag{3.82}$$

and therefore

$$p^{\hat{K}_1} \cdot D_1 = \frac{\partial f_i^{\hat{K}_1}}{\partial w_i} - p^{K_1} \cdot B_1 = \frac{\partial f_i^{\hat{K}_1}}{\partial w_i}. \tag{3.83}$$

As above, we see that

$$p^{K_2} \cdot A_2 = 0, \quad \omega_i^{K_2} = B_2 \cdot p^{\hat{K}_2} = 0, \tag{3.84}$$

so that $J \cap K_2 = \emptyset$.

Proceeding in the same way, we get

$$V^i = \begin{pmatrix} A_1 & & & 0 \\ & \ddots & & \\ & & A_f & \\ 0 & & & D_f \end{pmatrix}, \tag{3.85}$$

where D_f is an indecomposable matrix corresponding to the subset of commodities $\hat{K}_f$, $\hat{K}_f \cup K_f = \hat{K}_{f-1}$, $\hat{K}_f \cap K_f = \emptyset$ and

$$J \subset \hat{K}_f. \tag{3.86}$$

Arguing as in the construction of D_f, we see that D_f is a nonsingular matrix. Let $x \in \mathrm{Ker} V^i$. Then

$$D_f \cdot x^{\hat{K}_f} = 0. \tag{3.87}$$

Hence

$$x^{\hat{K}_f} = 0. \tag{3.88}$$

In view of (3.86), (3.67) now follows from (3.88), which concludes the proof. □

3.22 Since $f_i(p, \langle \omega_i, p \rangle) > 0$, the Perron-Frobenius theorem shows that, under the assumptions of Theorem 3.21, the system (3.21) has a solution if and only if V^i is a non-degenerate matrix. Hence, in the conditions of Theorem 3.21, Theorem 3.16 applies only to regular economies (when it reduces to Theorem 3.7). Thus to consider boundary economies with equilibrium gross substitutability we need to further generalize the definition of infinitesimal stability (other definitions from 3.2 and 3.15 can be generalized in a similar way).

Definition Let $\omega \in \overline{\Omega}^*$, and let p be an equilibrium price vector at ω. Suppose that (ω, p) reg. (i/K) $(1 \le i \le m,\ K \subset \{1, \dots, l\})$. Put

$$\Omega_{i/K} = \{\omega' \in \Omega_i \mid (\omega'_i)^k = \omega_i^k, k \notin K\}. \tag{3.89}$$

A tangent vector ϑ to W_i at the point (ω, p) will be called (i/K)-*actual* if ϑ is tangent to a smooth curve C in W_i such that

$$\pi_i(C) \subset \Omega_{i/K}. \tag{3.90}$$

The equilibrium corresponding to ω and p is called *actually infinitesimally unstable* with respect to the i-th participant and commodities from K ((ω, p) act.i.s. (i/K)) if for an arbitrary (i/K)-actual tangent vector ϑ to W_i at (ω, p) such that

$$d\pi_i(\vartheta) = \sum_{k \in K} \alpha_k \frac{\partial}{\partial \omega_i^k}, \quad \alpha_k \le 0,\ \sum_{k \in K} \alpha_k < 0, \tag{3.91}$$

one has that

$$\frac{\partial v_i^p}{\partial \vartheta} < 0. \tag{3.92}$$

When $K = J = \{j \mid \omega_i^j > 0\}$ we say that (ω, p) is actually infinitesimally stable with respect to the i-th participant ((ω, p) act.i.s. (i)).

We observe that if $\omega_i > 0$, then (ω, p) act.i.s. $(i) \Leftrightarrow (\omega, p)$ i.s. (i).

3.23 Theorem *Let $\omega \in \overline{\Omega}^*$, and suppose that for each economy $\omega' \in \Omega_{i/J}$ sufficiently close to ω there exists a unique (modulo a constant factor) equilibrium price vector and there is equilibrium gross substitutability. Suppose further that all the participants have normal demand. Then (ω, p) is actually infinitesimally stable with respect to the i-th participant.*

Proof. Let C be a smooth curve in W_i such that $\pi_i(C) \subset \Omega_{i/J}$, $(\omega, p) \in C$, and the tangent vectors at an arbitrary point $\xi \in C$ satisfy the condition (3.91). Proceeding as in 3.21, for each point $\xi \in C$ we construct a decomposition of the matrix $V^i(\xi)$. It is easy to see that if one chooses the curve C to be sufficiently small, then

$$\hat{K}_f(\xi) \supseteq \hat{K}_f(\omega, p). \tag{3.93}$$

We proceed by descending induction on $\hat{K}_f$. If $\hat{K}_f = \{1, \dots, l\}$, then the economy is regular and it suffices to argue as in the proof of Theorem 3.20 C).

Assuming that the claim of Theorem 3.23 is proved for all $\hat{K}_f \supseteq \hat{K}_f(\omega, p)$, $\hat{K}_f \neq \hat{K}_f(\omega, p)$, we show that (ω, p) act.i.s. (i). We consider the following two cases.

a) For all $\xi \in C$ sufficiently close to (ω, p) one has $\hat{K}_f(\xi) = \hat{K}_f(\omega, p) = K$. From (3.85) it follows that we may assume that

$$p^j(\xi) = p^j, \quad j \notin K. \tag{3.94}$$

Since the matrix D_f is non-degenerate and has all the properties of the matrix V^i, (3.92) immediately follows from Theorem 3.7 (cf. (3.63)).

b) Suppose that there exists a sequence $\xi_n \to (\omega, p)$ such that $\hat{K}_f(\xi_n) \supseteq \hat{K}_f(\omega, p)$, $\hat{K}_f(\xi_n) \neq \hat{K}_f(\omega, p)$. In view of the induction assumption, from the continuity it follows that $\frac{\partial v_i^p}{\partial \vartheta} \leq 0$. Now the strict inequality is established in the same way as in a). This concludes the proof. □

3.24 From the economic point of view, it is interesting to study the behavior of prices under throwing out of goods. However, in order to speak about *absolute* prices, one needs to introduce a suitable normalization. The most convenient normalization is $w_i \equiv 1$, where $1 \leq i \leq m$ is a fixed participant. For this normalization one has the following result.

Theorem *Let* $\omega \in \overline{\Omega}^*$.

a) *Suppose that* $\omega_i > 0$ *and* ω *i.s.* (i) (*resp.* ω *w.i.s.* (i)). *Then there exists a* (*local*) *throwing out for which the prices of all commodities* $j \in J$ *increase* (*do not decrease*).

b) *Suppose that* ω *has one of the properties listed in 3.20 and 3.23, and let* J *be defined by the formula* (3.56). *Then for each* (*local*) *throwing out the prices of all commodities* $j \in J$ *do not decrease* (*in the conditions of Theorem 3.20, the prices of* all *commodities do not decrease*).

Proof. a) In view of 2.7 and 21.8, under our normalization

$$d\pi_i = -V^i. \tag{3.95}$$

But by Theorem 3.7 (resp. 3.8)

$$(V^i)(f_i(p(\omega), \langle \omega, p(\omega) \rangle))_j > 0 \text{ (resp. } \geq 0), \quad j \in J, \tag{3.96}$$

and it suffices to perform throwing out along the (positive) vector $f_i(p(\omega), \langle \omega, p(\omega) \rangle)$.

b) In the conditions of Theorem 3.20 our claim follows from (3.95) and (3.63). In the conditions of Theorem 3.23 one needs to use a more detailed analysis performed in 3.21. □

4 Local and global stability

4.1 Throughout this section we assume that for a fixed initial endowment $\omega \in \overline{\Omega}^*$ and each initial endowment $\omega' \in \overline{\Omega}$, $\omega' \leq \omega$ there exists a unique (modulo a constant factor) equilibrium price vector (this assumption will be lifted elsewhere). We denote by $p(\omega)$ the (normalized) equilibrium price vector corresponding to ω.

Definition An economy ω is called *globally* (resp. *locally*) *stable* with respect to the i-th participant if for each economy $\omega' \in \overline{\Omega}_i$ (resp. for each each economy $\omega' \in \overline{\Omega}_i$ sufficiently close to ω), $\omega' \leq \omega$ one has

$$v_i(\omega') < v_i(\omega), \tag{4.1}$$

where

$$v_i(\omega) = u_i(f_i(p(\omega), \langle \omega_i, p(\omega) \rangle)) \tag{4.2}$$

(cf. (3.2)).

In a similar way one can introduce local and global counterparts of the other definitions from 3.2.

The connection between the notions of infinitesimal, local and global stability is revealed by the following theorems.

4.2. Theorem *If ω is globally stable with respect to the i-th participant, then ω is locally stable with respect to the same participant. Conversely, suppose that for each economy $\omega' \in \overline{\Omega}_i$, $\omega' \leq \omega$ the economy ω' is locally stable with respect to the i-th participant. Then ω is globally stable.*

The proof is obvious.

4.3 Theorem *Suppose that ω reg. (i/J) and $(\omega, p(\omega))$ i.s. (i/J), where $J = \{j \mid \omega_i^j > 0\}$. Then ω is locally stable with respect to the i-th participant.*

The proof is obvious.

4.4 Corollary *Suppose that ω' reg. (i/J) for each $\omega' \in \overline{\Omega}_i$, $\omega' \leq \omega$ and that $(\omega', p(\omega'))$ i.s. (i/J). Then ω is globally stable.*

4.5 Theorem *Let $\omega \in \overline{\Omega}^*$ be an economy which is globally stable with respect to the i-th participant and such that $\omega_i > 0$, $\sum_{k \neq i} \omega_k > 0$. Suppose that ω is regular and infinitesimally stable with respect to the i-th participant. Then each economy $\tilde{\omega}$ sufficiently close to ω is globally stable with respect to the i-th participant.*

Proof. As was observed in 3.3, there exists a neighborhood $U \ni \omega$ such that each economy $\omega' \in U$ satisfies the assumptions of the theorem except, possibly, global stability. Let $U' \subset U$ be a smaller neighborhood of ω. The function v_i attains maximum on the compact $\{\omega' \in \overline{\Omega}_i \mid \omega'_i \leq \omega_i,\ \omega' \notin U'\}$; let $\bar{v}$ be this maximum. Let

$$v_i(\omega) - \bar{v} = \varepsilon > 0. \tag{4.3}$$

Since the function v_i is continuous on the compact

$$K = \{\omega' \in \overline{\Omega} \mid \omega' \in \overline{\Omega}_i(\bar{\omega}), \quad \omega'_i \leq \bar{\omega}_i\ \exists \bar{\omega} \in \bar{U}\}, \tag{4.4}$$

v_i is uniformly continuous on this compact. Choose a constant $\delta > 0$ so that for $x, y \in K$, $\|x - y\| < \delta$ one has

$$\|v_i(x) - v_i(y)\| < \frac{\varepsilon}{2} \tag{4.5}$$

and

$$\delta < \operatorname{dist}(\bar{U}', \partial \bar{U}). \tag{4.6}$$

We claim that for

$$\tilde{\omega} \in U', \quad \|\tilde{\omega} - \omega\| < \delta, \tag{4.7}$$

$\tilde{\omega}$ is a globally stable economy. Suppose the converse, and let $\omega' \in \overline{\Omega}_i(\tilde{\omega})$,

$$\omega'_i \leq \tilde{\omega}_i, \ v_i(\omega') \geq v_i(\tilde{\omega}). \tag{4.8}$$

Since each economy from U is infinitesimally stable with respect to the i-th participant, an economy ω' satisfying (4.8) cannot lie in U.

Consider an economy $\omega'' \in \overline{\Omega}_i(\omega)$ such that

$$(\omega''_i)^j = \max\{(\omega'_i + \omega_i - \tilde{\omega}_i)^j, 0\}, \quad j = 1, \ldots, l. \tag{4.9}$$

Then $\|\omega' - \omega''\| < \delta$, and therefore

$$\|v_i(\omega') - v_i(\omega'')\| < \frac{\varepsilon}{2}, \quad \omega'' \notin U'. \tag{4.10}$$

Since $\omega'' \leq \omega$,

$$v_i(\omega'') < \bar{v} = v_i(\omega) - \varepsilon. \tag{4.11}$$

From (4.10), (4.8), (4.7) and (4.5) it follows that

$$v_i(\omega'') > v_i(\omega'') - \frac{\varepsilon}{2} \geq v_i(\tilde{\omega}) - \frac{\varepsilon}{2} > v_i(\omega) - \varepsilon, \tag{4.12}$$

which contradicts (4.11). This contradiction completes the proof. □

We notice that, under suitable conditions, Theorem 4.5 can be generalized to the case of boundary economies.

4.6 Theorem *Each economy from a neighborhood of the Pareto subset (i.e. the set of Pareto optimal allocations) is globally stable with respect to each of the participants.*

Proof. Since a Pareto optimal allocation $\omega \in \overline{\Omega}^*$ is regular, Corollary 3.11 and Theorem 4.5 show that it is sufficient to prove global stability with respect to each of the participants of ω itself.

Let $1 \leq i \leq m$, and let $\omega' \in \overline{\Omega}_i$, $\omega'_i \leq \omega_i$. Suppose that

$$v_i(\omega') \geq v_i(\omega). \tag{4.13}$$

Consider an allocation $\bar{\omega}$ such that

$$\begin{cases} \bar{\omega}_k = f_k(p(\omega'), \langle \omega'_k, p(\omega') \rangle), & k \neq i, \\ \bar{\omega}_i = f_i(p(\omega'), \langle \omega'_i, p(\omega') \rangle) + (\omega_i - \omega'_i). \end{cases} \tag{4.14}$$

Then

$$\sum_{k=1}^m \bar{\omega}_k = (\omega_i - \omega'_i) + \sum_{k=1}^m \omega'_k = \omega_i + \sum_{k \neq i} \omega_k = \sum_{k=1}^m \omega_k, \tag{4.15}$$

where

$$u_k(\bar{\omega}_k) \geq u_k(\omega'_k) = u_k(\omega_k), \quad k \neq i \tag{4.16}$$

and, in view of strict monotonicity of u_i and condition (4.13),

$$u_i(\bar{\omega}_i) > v_i(\omega') \geq v_i(\omega) = u_i(\omega_i). \tag{4.17}$$

Relations (4.15)–(4.17) show that $\bar{\omega}$ is preferred to ω, contrary to the assumption that ω is Pareto optimal. Hence $v_i(\omega') < v_i(\omega)$ and ω is globally stable, which proves the theorem. □

4.7 Theorem *Let $\omega \in \overline{\Omega}^*$ be a (not necessarily stable) economy. Suppose that the i-th participant has positive initial endowments (i.e. $\omega_i > 0$) and normal demand. Then for each economy $\omega' \in \overline{\Omega}_i$ such that $\omega'_i < \omega_i$ there exists an economy $\omega'' \in \overline{\Omega}_i$ such that $\omega'_i \le \omega''_i < \omega_i$ and $v_i(\omega') < v_i(\omega'')$. In other words, the i-th participant does not gain from throwing out all commodities.*

Proof. By Theorem 3.6, the function v_i is differentiable at the point ω' along the vector

$$\vartheta = \sum_{j=1}^{l} \frac{\partial f_i^j}{\partial \omega_i}(p(\omega'), \langle \omega', p(\omega')\rangle)\frac{\partial}{\partial \omega_i^j}, \tag{4.18}$$

where

$$\vartheta \ge 0, \quad \frac{\partial v_i}{\partial \vartheta} > 0. \tag{4.19}$$

Consider an economy $\omega''_\varepsilon \in \overline{\Omega}_i$ such that

$$(\omega''_\varepsilon)_i = \omega'_i + \varepsilon \frac{\partial f_i}{\partial w_i}(p(\omega'), \langle \omega', p(\omega')\rangle), \tag{4.20}$$

where ε is a positive number. Since demand is normal, one has $\omega''_\varepsilon \ge \omega'$. Since $\omega'_i < \omega_i$, for small ε one has $(\omega''_\varepsilon)_i < \omega_i$. Finally, from (4.19) it follows that $v_i(\omega''_\varepsilon) > v_i(\omega')$ provided that ε is small enough. This concludes the proof. □

4.8 Theorem *The set of economies $\omega \in \overline{\Omega}^*$ which are weakly globally stable with respect to the i-th participant is closed in $\overline{\Omega}^*$ (we recall that the definition of weak global stability is obtained from the definition of global stability given in 4.1 by replacing strict inequality* (4.1) *by a non-strict one).*

Proof. Suppose that the assertion of the theorem is false. Then there exist an economy $\omega \in \overline{\Omega}^*$ and a sequence of economies $\omega(n)$ weakly globally stable with respect to the i-th participant such that

$$\lim_{n\to\infty} \omega(n) = \omega \tag{4.21}$$

and an economy

$$\omega' \in \overline{\Omega}_i(\omega), \quad \omega'_i \le \omega_i \tag{4.22}$$

such that

$$v_i(\omega') > v)_i(\omega). \tag{4.23}$$

We define an economy $\omega''(n) \in \overline{\Omega}_i(\omega(n))$ by putting

$$(\omega_i''(n))^j = \max\{(\omega_i' - \omega_i + \omega_i(n))^j, 0\}, \quad j = 1, \ldots, l. \tag{4.24}$$

It is clear that

$$\omega_i''(n) \leq \omega_i(n), \tag{4.25}$$

and by our assumptions

$$v_i(\omega''(n)) \leq v_i(\omega(n)). \tag{4.26}$$

But it is easy to see that

$$\lim_{n\to\infty} \omega''(n) = \omega'. \tag{4.27}$$

From (4.21), (4.26), (4.27) and the continuity of the function v_i it follows that

$$v_i(\omega') \leq v_i(\omega), \tag{4.28}$$

contrary to (4.23). This contradiction completes the proof. □

4.9 We proceed with extending 4.4 to the case of not necessarily regular economies.

Theorem *Let $K \subset \overline{\Omega}$ be a nowhere dense closed subset of measure zero containing the set of all critical economies. Suppose that each economy $\omega \in \overline{\Omega} \setminus K$ is weakly infinitesimally stable with respect to the i-th participant. Then each economy $\omega \in \overline{\Omega}$ is weakly globally stable with respect to the i-th participant.*

Proof. The proof is easy when $f(p; w_1, \ldots, w_m)$ is an analytic function. In general, it is sufficient to verify that if $\omega \in \overline{\Omega}$, $\omega' \in \overline{\Omega}_i$ and $\omega_i' \leq \omega_i$, then $v_i(\omega') \leq v_i(\omega)$.

Let $\alpha = \alpha(\omega, \omega')$ be the constant vector field on $\overline{\Omega}_i$ consisting of vectors of length one whose integral curves are lines parallel to the line joining ω_i with ω_i' with orientation from ω_i to ω_i'. Let ξ be a continuous vector field on $\overline{\Omega}_i$ consisting of vectors of length one collinear at each point to the vector field $\frac{\partial f_i}{\partial w_i}$ (this last vector field is non-vanishing in view of (2.27)). Further, let δ be a positive number. We denote by K_δ a union of balls covering K the sum of whose volumes does not exceed δ. From the existence of partition of unity it follows that there exists a continuous vector field $\beta = \beta(\omega)$ on $\overline{\Omega}_i$ consisting of vectors of length one such that

$$\begin{array}{ll} \beta(\tilde{\omega}) = \alpha(\tilde{\omega}), & \tilde{\omega} \notin K_\delta; \\ \beta(\tilde{\omega}) = \xi(\tilde{\omega}), & \tilde{\omega} \in K. \end{array} \tag{4.29}$$

It is worthwhile to observe that, although the function v_i need not be differentiable (at the points of K), in view of (3.18), v_i is differentiable along the vector field β. We choose $\delta = \delta(\varepsilon)$ so small that there exists a point $\tilde{\omega} \in \overline{\Omega}_i$ such that

$$\|\tilde{\omega} - \omega\| < \varepsilon \tag{4.30}$$

and that for the integral curve $C = C(\tilde{\omega}, \delta)$ of the vector field $\beta(\delta)$ passing through $\tilde{\omega}$ one has

$$\mu(C \cap K_\delta) < \varepsilon \tag{4.31}$$

(where μ is the standard measure on the curve C).

Let T be the time required for the current corresponding to the vector field α to run through the interval $[\omega_i, \omega_i']$, and let $\tilde{\omega}' = \tilde{\omega}'(\tilde{\omega}, \delta)$ be the end of the interval of the integral curve $C(\tilde{\omega}, \delta)$ run through during the time T. It is clear that

$$\lim_{\varepsilon \to 0} \tilde{\omega}' = \omega'. \tag{4.32}$$

One has

$$\begin{array}{rcl} v_i(\tilde{\omega}') - v_i(\tilde{\omega}) & = & \int_0^T (v_i)'_\beta dt \\ & = & \int_{C(t) \notin K_\delta} (v_i)'_\beta dt + \int_{C(t) \in K_\delta} (v_i)'_\beta dt \\ & = & \int_{C(t) \notin K_\delta} (v_i)'_\alpha dt + \int_{C(t) \in K_\delta} (v_i)'_\beta dt. \end{array}$$

Since the economies outside of K are weakly infinitesimally stable, the first summand in (4) is non-positive. On the other hand,

$$\int_{C(t) \in K_\delta} (v_i)'_\beta dt \leq \left| \int_{C(t) \in K_\delta} (v_i)'_\xi dt \right|. \tag{4.33}$$

Since the integrand in the right hand part of (4.33) does not depend on ε, one has

$$v_i(\tilde{\omega}') - v_i(\tilde{\omega}) \leq \varepsilon M, \tag{4.34}$$

where M is a constant which does not depend on ε (this argument can be further simplified if the i-th participant has normal demand).

Letting $\varepsilon \to 0$ and using (4.30), (4.32) and the continuity of the function v_i, we see that

$$v_i(\omega') \leq v_i(\omega). \tag{4.35}$$

This proves the theorem. □

4.10 Remark If in the statement of Theorem 4.9 one replaces the assumption of weak infinitesimal stability by the assumption of infinitesimal stability, then from our proof it follows that all economies from $\overline{\Omega} \setminus K$ are globally stable (this result can be extended to the boundary economies by replacing the assumptions of regularity and infinitesimal stability by the assumptions of (i/J)-regularity and (i/J)-infinitesimal stability (cf. 4.4)).

4.11 We turn to a study of systems with gross substitutability.

Theorem *Let $\omega \in \overline{\Omega}^*$, and suppose that for each $\omega' \in \overline{\Omega}$, $\omega' \leq \omega$ one has equilibrium gross substitutability and all the participants have normal demand. Then ω is globally stable with respect to each of the participants.*

Proof. In the case when at least one of the assumptions a)–c) of Theorem 3.20 is satisfied for at least one of the participants, Theorem 4.11 follows from Theorem 3.20 and Corollary 4.4 (we observe that *weak* global stability of ω follows from an obvious variant of Theorem 4.9).

To prove Theorem 4.11 in the general case we use the fact that, according to Theorem 3.23, $(\omega, p(\omega))$ act.i.s. $(1, \dots, m)$. The global stability is deduced from the actual infinitesimal stability in the same way as it is deduced from the infinitesimal stability in Corollary 4.4. □

We remark that the assumption of equilibrium gross substitutability is satisfied, in particular, in the case of systems with global gross substitutability (i.e. in the case when the inequality (3.57) holds for *all* $p \in I\!R^l_+$).

4.12 Theorem *Let $\omega \in \overline{\Omega}^*$, and suppose that for each $\omega' \in \overline{\Omega}_i$ one has equilibrium gross substitutability and all the participants have normal demand. Then, under the normalization $w_i \equiv 1$, $p^j(\omega') \geq p^j(\omega)$ for all $j \in J$. If, furthermore, ω satisfies one of the conditions of Theorem 3.20, then $p^j(\omega') \geq p^j(\omega)$ for all j.*

Proof. Theorem 4.12 immediately follows from Theorem 3.24. □

Acknowledgement Translation from Russian, pp. 72–106 in "Methods of the theory of extremal problems in economics" (edited by V. L. Levin), Nauka, 1981.

References

1. Aumann, R.J., and B. Peleg (1975): "A note on Gale's example," *Journal of Mathematical Economics*, 1, 209–211.

2. Balasko, Y. (1975): "Some results on uniqueness and on stability of equilibrium in general equilibrium theory, " *Journal of Mathematical Economics*, 2, 95–118.

3. Debreu, G. (1970): "Economics with a finite set of equilibria," *Econometrica*, 38, 387–392.

4. Debreu, G. (1972): "Smooth preferences," *Econometrica*, 40, 603–616.

5. Dierker, E. (1974): *Topological methods in Walrasian economics.* Lecture Notes in Economics and Mathematical Systems, Springer-Verlag, Vol. 92, Berlin-Heidelberg-New York.

6. Efimov, B.A., and A.S. Shapovalov (1979): "Unstable economies and coalitions," (in Russian) in: Studies in Mathematical Economics and Related Topics, Moscow State University, Moscow, 103–113.

7. Gale, D. (1974): "Exchange equilibrium and coalitions; an example," *Journal of Mathematical Economics*, 1, 63-66.

8. Intriligator, M. (1971): *Mathematical Optimization and Economic Theory.* Prentice Hall.

9. Nikaido, H. (1997): *Convex Structures and Economic Theory.* Academic Press.

10. Polterovich, V.M., and V.A. Spivak (1978): *Coalition Stability of Economic Equilibrium* (in Russian). CEMI, Moscow.

Chapter 12

An algorithmic approach for searching an equilibrium in fixed budget exchange models

V.I. Shmyrëv

Abstract: In this paper we consider an approach for searching an equilibrium in a linear fixed budgets exchange model and its "antipode" the so-called cooperation model. This approach, which may be characterized as a polyhedral complementarity one, leads to the consideration of piecewise constant point-to-set mappings of the unit price simplex into itself. The fixed points of these mappings yield the equilibrium prices. Some finite algorithms are proposed.

Key words: exchange model, equilibrium, transportation problem, monotonicity, fixed point.

1 Introduction

In this paper we consider finite algorithms for determining equilibrium prices in a linear fixed budgets exchange model.

For the general case of the linear exchange model the algorithm proposed by B. Eaves (1976) is well-known. Eaves (1976) reduces the problem to a linear complementarity one and then applies Lemke's algorithm (1965) to solve the problem. This reduction is complicated by the fact that the arising complementarity problem is of a relatively large dimension, being equal to the product of the number of agents and the number of commodities.

The approach considered in this paper leads to a more compact scheme, that may be characterized as polyhedral complementarity. This approach was proposed by the author in (1983) for the general case of a linear exchange model. The linearity of the goal functions of the agents yields a subdivision (generated by the model) of the price simplex into the agent's preference stable domains. To each of these domains there corresponds a set of price vectors, for which balanced exchange of goods is possible according to the agents' preferences. The fixed points of this constructed point-to-set mapping yield the equilibrium price vectors of the model (Shmyrëv, 1981b).

The mentioned approach makes it possible to clarify some quality aspects of the problem, namely that the constructed point-to-set mapping has some monotonicity property. This property can be characterized as follows: the generated polyhedral complementarity problem locally reduces to the linear complementarity one with positive principal minors of the constraint matrix (P-class: Lemke, 1965). To make the revealed property more obvious, an "antipodal model" to the exchange model is considered, which we call the cooperation model. In this model the utility functions are replaced by "disutility functions", which the agents want to minimize under the given constraints. In this case the mentioned monotonicity property of the point-to-set mapping changes into the opposite: in case of an exchange model the mapping has a decreasing property, in case of an cooperation model the mapping is increasing.

In this paper we consider the exploitation of the mentioned approach for the fixed budgets model. For those models the constructed mapping proves to be potential and the monotonicity property appears to be logarithmic. The proposed finite algorithms for searching an equilibrium are similar to the author's algorithms (1981a, 1982) for searching fixed points of a monotone piecewise constant point-to-set mapping in $I\!R^n$.

The paper is organized as follows. Section 2 contains the general description of the considered models. In Section 3 theorems characterizing the equilibrium points as stationary points are given. The proposed finite algorithms are considered in Section 4. Section 5 discusses some generalizations.

2 The models

In the well-known economic model of an exchange economy there are n goods (commodities) and m consumers (agents). Each consumer i possesses a vector of initial endowments $d^i = (d^i_1, \ldots, d^i_n)$ and must choose a consumption vector $x^i = (x^i_1, \ldots, x^i_n)$ maximizing his utility function $U_i(x^i)$. Here d^i_j is the quantity of good j in possession of consumer i, and x^i_j is the consumption of good j by consumer i. So the supply of consumer i is d^i and his demand is x^i.

The exchange is realized with respect to nonnegative prices p_j, forming price vector $p \in I\!R^n_+$. Thus each consumer i solves the mathematical program

$$\begin{aligned} U_i(x^i) &\to \max! \\ (p, x^i) &\leq (p, d^i), \\ x^i &\geq 0, \end{aligned}$$

where (x, y) denotes the inner product of $x,\ y \in I\!R^n$. For each i, let $\bar{x}^i$ be an x^i that solves this program. Clearly, it will hold that $(p, \bar{x}^i) = (p, d^i)$.

A price vector $\bar{p} \neq 0$ is an equilibrium price vector if there exist solutions $\bar{x}^i$, $i = 1, \ldots, m$, for the individual (utility) optimization problems such that $\sum \bar{x}^i = \sum d^i$.

In the following we normalize the initial endowment of each commodity to one, i.e., $\sum_i d^i = (1, \ldots, 1) = \theta \in I\!R^n$. Also the sum of the prices p_j is normalized to one, restricting the price vector p to be in the unit simplex denoted by σ:

$$\sigma = \left\{ p \in I\!R^n_+ \,|(p, \theta) = 1 \right\}.$$

In the linear exchange model all utility functions are linear: $U_i(x^i) = (c^i, x^i)$ with c^i a nonnegative vector. In the special case that $d^i = \lambda_i \sum\limits_{k=1}^{m} d^k$ with $\lambda_i \geq 0$, $\sum\limits_{i=1}^{m} \lambda_i = 1$, we have $(p, d^i) = \lambda_i$ for each $p \in \sigma$, and such a model is called the linear fixed budgets exchange model.

For the exchange model (even for the general case with production) T. Negishi (1960) has shown that there exists a social welfare function (a linear combination of the utility functions $U_i(x^i)$) such that the equilibrium problem reduces to maximizing this function under the market equality constraints. In this paper we restrict ourselves to the linear fixed budgets exchange model. For this case D. Gale (1960) has shown that the social welfare function $\gamma(x^1, \ldots, x^n) = \prod\limits_{i=1}^{m} (c^i, x^i)^{\lambda_i}$ satisfies. Our approach bases on using another optimization problem, different from that by T. Negishi and D. Gale.

Besides the exchange model we also consider its "antipode" , which we call a cooperation model. The difference is that instead of maximizing the goal function (c^i, x^i), each agent now wants to minimize this function (under the previous assumption that $c^i \geq 0$). The goal function can now be seen as a "disutility function": more consumption gives lower utility. One can imagine the following interpretation of this cooperation model. The set of all agents has to perform n duties. Initially agent i has to perform the jth duty in quantity d^i_j, i.e. the vector d^i is the vector of duties assigned to agent i. However, the agents have the possibility to exchange the duties in order to minimize their disutilities. In contrast to the exchange model, in this case agent i has to pay (p, d^i) for his vector of duties d^i and gets (p, x^i) if he accepts the vector of duties x^i. So, given the vector p of prices, the optimization problem of the ith agent becomes:

$$\begin{gathered}(c^i, x^i) \to \min!\\ (p, x^i) \geq (p, d^i),\\ x^i \geq 0.\end{gathered}$$

Again, at an optimal solution $x^i = \bar{x}^i$ it holds that $(p, \bar{x}^i) = (p, d^i)$ and in the case of the fixed budgets model one has as before that $(p, d^i) = \lambda_i$. The definition of equilibrium price vector remains the same as in exchange model, but in this case "optimization" means "minimization".

For simplicity and uniformity we shall use opt instead max and min in both the exchange and cooperation models. So, for the fixed budget models the ith agent has in both cases to solve the problem:

$$\begin{gathered}(c^i, x^i) \to \text{opt!}\\ (p, x^i) = \lambda_i,\\ x^i \geq 0.\end{gathered}$$

3 Stationary point characterizations of equilibrium points

For simplicity we shall assume $c^i > 0, i \in I$. In line with Shmyrëv (1981b), we associate to each exchange or cooperation model the following parametric transportation problem:

$$\sum_{i\in I}\sum_{j\in J} z_{ij} \ln c_{ij} \to \text{opt!} \tag{3.1}$$

$$\sum_{j\in J} z_{ij} = \lambda_i, \qquad i \in I, \tag{3.2}$$

$$\sum_{i\in I} z_{ij} = p_j, \qquad j \in J, \tag{3.3}$$

$$z_{ij} \geq 0, \quad i \in I, \quad j \in J. \tag{3.4}$$

Here c_{ij} are components of the vectors c^i, $i \in I$, and the p_j are components of the price vector $p \in \sigma$. The prices p_j are the parameters of this problem.

The system of restrictions of the dual problem in the case of the exchange model is

$$u_i + v_j \geq \ln c_{ij}, \quad i \in I, \ j \in J, \tag{3.5}$$

whereas for the cooperation model the symbol of the inequality should be changed to the opposite. The problem (3.1)–(3.4) is solvable for each $p \in \sigma$. Let $W(p) \subset I\!R^{m+n}$ be the set of vectors $w = (u_1, \ldots, u_m, v_1, \ldots, v_n)$ defining an optimal solution of the dual problem for a given $p \in \sigma$, and $V(p) \subset I\!R^n$ be the projection of this set to the subspace of variables $v_1, \ldots, v_n$. Denote by $\Phi(p)$ the set

$$\Phi(p) = \{e^v = (e^{v_1}, \ldots, e^{v_n}) \mid v = (v_1, \ldots, v_n) \in V(p)\} \cap \sigma.$$

Clearly, for each $p \in \sigma$, the set $\Phi(p)$ lies in the relative interior of the simplex σ, i. e., $\Phi(p) \subset \sigma^\circ$. Thus, the point-to-set mapping $\Phi : \sigma \to 2^{\sigma^\circ}$ is defined.

Theorem 3.1 *A vector $p \in \sigma$ is an equilibrium price vector if and only if it is a fixed point of the mapping Φ.*

Proof. Consider a vector $\bar{p} \in \Phi(\bar{p})$ and an optimal solution $z_{ij} = \bar{z}_{ij}$, $(i,j) \in I \times J$, of the transportation problem for $p = \bar{p}$. Then $\bar{p} > 0$, and due to (3.3), the vectors $\bar{x}^i$ whose components are $\bar{x}_{ij} = (\bar{z}_{ij}/\bar{p}_j) \geq 0$ satisfy the condition $\sum_{i\in I} \bar{x}^i = \theta$.

Let us show that the vector $\bar{x}^i$ solves the problem of the ith agent. It follows from (3.2) that $\bar{x}^i$ is a feasible vector in this problem: $(\bar{p}, \bar{x}^i) = \lambda_i$. Let us verify its optimality. We restrict the exposition to the exchange model (the case of the cooperation model goes in a similar way). It follows from $\bar{p} \in \Phi(\bar{p})$ that $\ln \bar{p} = (\ln \bar{p}_1, \ldots, \ln \bar{p}_n) \in V(\bar{p})$. This means that there exists a vector $\bar{u} = (\bar{u}_1, \ldots, \bar{u}_m)$, which together with $\bar{v} = \ln \bar{p}$ forms an optimal solution to the dual of the transportation problem: $\bar{w} = (\bar{u}_1, \ldots, \bar{u}_m, \bar{v}_1, \ldots, \bar{v}_n) \in W(p)$. Thus, $u_i = \bar{u}_i$ and $v_j = \bar{v}_j$ satisfy (3.5), and if $z_{ij} > 0$, then the respective inequality of (3.5) turns to equality. Introducing $\bar{y}_i = e^{\bar{u}_i}$, we obtain that

$$\bar{y}_i \bar{p}_j \geq c_{ij}, \quad (i,j) \in I \times J, \tag{3.6}$$

and the equality holds if $\bar{z}_{ij} > 0$, i. e., if $\bar{x}_{ij} > 0$. So, the vector $\bar{x}^i$ is optimal in the problem of the ith agent. This proves that each fixed point of Φ determines an equilibrium price vector of the model.

To show the opposite, consider $p = \bar{p} \in \sigma$ and suppose that the vectors $\bar{x}^i$, $i \in I$, are optimal in the respective agents' problems and $\sum_{i \in I} \bar{x}^i = \theta$. Clearly, this can only be true if $\bar{p} > 0$. Indeed, if we consider the exchange model with some component of $\bar{p}$ equal to zero, then the agents' problems cannot have optimal vectors due to the positivity of the vectors c^i. If we consider the cooperation model, then $\bar{p}_{j_\circ} = 0$ implies $\bar{x}_{ij_\circ} = 0 \ \forall i \in I$, and thus $0 = \sum_{i \in I} x_{ij_\circ} \neq 1$.

The remaining arguments repeat those of the first part of the proof taken in the reverse order. Again, we only consider the exchange model. Since $\bar{x}^i$ is optimal for the problem of the ith agent, there exists some $\bar{y}_i$ such that the system (3.6) holds, and if $\bar{x}_{ij} > 0$, then the respective inequality of the system turns to the equality. Clearly $\bar{y}_i > 0$. Taking $\bar{z}_{ij} = \bar{p}_j \bar{x}_{ij}$ and $\bar{u}_i = \ln \bar{y}_i$, $\bar{v}_j = \ln p_j$, it can be easily seen that we obtain an optimal solution of the problem (3.1)–(3.4) and to its dual problem respectively. Thus, $\ln \bar{p} \in V(\bar{p})$, i. e., $\bar{p} \in \Phi(\bar{p})$. □

To search for fixed points of the mapping Φ, we shall use the fact that the mapping $V : \sigma \to 2^{I\!R^n}$ generating Φ is potential in the following sense. Let $f : \sigma \to I\!R^1$ be the function which for each $p \in \sigma$ is equal to the optimal value of the goal function of (3.1)–(3.4). For the cooperation model, the function f is convex and

$$\partial f(p) = V(p), \tag{3.7}$$

where $\partial f(p)$ is the subdifferential of f at the point p. For the exchange model, the function f is concave, and as it can be easily seen, for the convex function $g = -f$ we shall have $-\partial g(p) = V(p)$. For the sake of simplicity and uniformity, in this case it is appropriate to denote the subdifferential of f at the point p by $\partial f(p)$, defining it by $\partial f(p) = -\partial g(p)$. So, equality (3.7) will hold for the exchange model as well. Now the condition for $p \in \sigma^\circ$ to be an equilibrium vector can be written as $\ln p \in \partial f(p)$.

Let $h(p)$ be the function defined by $h(p) = (p, \ln p)$ for $p \in \sigma^\circ$ and continuously extended to the boundary of σ by zero values (the well known entropy function with the opposite sign). Consider the function

$$\varphi(p) = h(p) - f(p)$$

on σ. For the exchange model, the function φ is convex, however for the cooperation model φ is the difference of two convex functions having the same efficient domain $\mathrm{dom}(h) = \mathrm{dom}(f) = \sigma$.

For each $p \in \sigma^\circ$, the function φ is differentiable on each direction q which does not lead out from σ° by small shifts. Using the known formula for the directional derivative of a convex function, we obtain the following representation for the derivative $\varphi'(p, q)$:

$$\varphi'(p, q) = \mathrm{opt}_{v \in \partial f(p)}(q, \ln p - v). \tag{3.8}$$

Clearly, this equality implies that the relation $\ln p^\circ \in \partial f(p^\circ)$ is equivalent to the condition that for each $q = (q_1, q_2, \ldots, q_n) \neq 0$, $\sum_{j \in J} q_j = 0$, the inequality holds $\varphi'(p^\circ, q)\varphi'(p^\circ, -q) \geq 0$. The point p° for which this condition holds is naturally called a *stationary* point of the function φ.

Summarizing the preceding, we can formulate the following theorem:

Theorem 3.2 *A vector $p \in \sigma$ is an equilibrium price vector if and only if it is a stationary point of the function φ.*

Using conjugate functions, we obtain the dual version of this statement. Let f^* be the function conjugate to f. If f is concave (i.e., if the exchange model is considered), we by definition assume that $f^*(y) = -(-f)^*(-y)$, where $(-f)^*$ is the usual conjugate to the convex function $g = -f$.

Theorem 3.3 *A vector $p \in \sigma$ is an equilibrium price vector if and only if it is a stationary point of the function $\psi(p) = -f^*(\ln p)$ on σ°.*

Proof. Let p° be an equilibrium point, i. e., $\ln p^\circ \in \partial f(p^\circ)$. Since $\partial h(p^\circ)$ is the set of points of the form $x = \ln p^\circ + \lambda\theta$, $\lambda \in I\!R^1$, it follows that the condition $\ln p^\circ \in \partial g(p^\circ)$ can be rewritten as $\ln p^\circ \in \partial h(p^\circ) \cap \partial f(p^\circ)$. This means that $p^\circ \in \partial h^*(\ln p^\circ) \cap \partial f^*(\ln p^\circ)$. But $h^*(x) = \ln \sum_{j \in J} e^{x_j}$, and $\partial h^*(x)$ contains only one vector $q(x) = (q_1(x), \ldots, q_n(x))$ with $q_i(x) = e^{x_i} / \sum_{j \in J} e^{x_j}$ for each $x \in I\!R^n$ (here x_j are components of x). Thus, $q(\ln p^\circ) \in \partial f^*(\ln p^\circ)$; this means that $x^\circ = \ln p^\circ$ is a stationary point of the function $g(x) = h^*(x) - f^*(x)$. Changing the variables $x = \ln p$, $p > 0$, we obtain that p° is a stationary point of $\bar{\psi}(p) = g(\ln p)$ on the set $\mathrm{Int} I\!R^n_+$ and thus on σ°. It remains to take into account that we have $h^*(\ln p) = \ln \sum_{j \in J} p_j = \ln 1 = 0$ on σ°, hence $\bar{\psi}(p) = \psi(p)$. Thus, p° is a stationary point of $\psi(p)$ on σ°, which proves the statement.

The reverse can be proved analogously. We should just take into account that the function g takes constant values at lines whose direction vector is $\theta = (1, \ldots, 1)$, and thus a point p° is stationary for ψ on σ° if and only if it is stationary for $\bar{\psi}$ on $\mathrm{Int}\mathbb{R}^n_+$. □

4 Algorithms

The theorems obtained in the previous section characterizing the equilibrium points make it possible to propose and justify different algorithms to find an equilibrium. Let us consider the iteration method for the cooperation model given by the relation $p^{k+1} \in \Phi(p^k)$, i. e., for a point $p^k \in \sigma$, we choose as p^{k+1} any of the points of $\Phi(p^k)$.

Theorem 4.1 *Applying the iteration method to the cooperation model gives an equilibrium point in a finite number of steps.*

Proof. First, let us prove the following property of the function ψ for the cooperation model: for $q \in \Phi(p)$, the inequality

$$\psi(q) - \psi(p) \geq (p, \ln p - \ln q). \tag{4.1}$$

holds. Indeed, the condition $q \in \Phi(p)$ implies $\ln q \in \partial f(p)$, i. e., $p \in \partial f^*(\ln q)$. The function f^* for the cooperation model is convex, and thus

$$f^*(\ln p) \geq f^*(\ln q) + (p, \ln p - \ln q), \tag{4.2}$$

which is equivalent to (4.1).

Now let us apply (4.1) to $q = p^{k+1}$ and $p = p^k$. Note that due to a known property of the entropy, the right part of this inequality is non-negative for all $p, q \in \sigma$ and is equal to zero only for $q = p$. Thus, when $p^{k+1} \neq p^k$ we have $\psi(p^{k+1}) - \psi(p^k) > 0$. Moreover,

$$\psi(p^{k+1}) > \psi(p) \quad \forall p \in \Phi^{-1}(p^{k+1}).$$

Consequently, for $s > k+1$, the current point p^s of the process cannot be into the set $\Phi^{-1}(p^{k+1})$ where p^k is. It remains to take into account that since the set of feasible solutions of the problem dual to (3.1)–(3.4) is polyhedral, there is a finite number of different sets among $\Phi^{-1}(p)$, $p \in \sigma^\circ$. This fact implies that the iteration process is finite. □

Note that when using the iteration method, at each step we can use the dual method of successive improvement to solve the transportation problem (3.1)–(3.4) for a fixed $p = p^{k+1}$. Next we consider another method based on the same procedure and allowing us to find an equilibrium state in the cooperation model.

Assume that at some step of the process we know some dual feasible basis set $\mathcal{B}_k$ of the problem (3.1)–(3.4). From the system

$$u_i + v_j = \ln c_{ij}, \quad (i,j) \in \mathcal{B}_k.$$

we get u_i and v_j ap to a constant added to u_i and subtracted from v_j. Thus, the vector $p^k \in \sigma^\circ$ is uniquely determined:

$$p_j^k = e^{v_j} \Big/ \sum_{l=1}^{n} e^{v_l}, \quad j \in J.$$

Consider the system of constraints (3.2)–(3.3) and add to it the condition $z_{ij} = 0$, $(i,j) \notin \mathcal{B}_k$. The system obtained has the unique solution $z_{ij} = z_{ij}^k$. If it is non-negative, then it means that p^k is an equilibrium price vector of the model. If $z_{i_\circ j_\circ} < 0$ for some pair $(i_\circ, j_\circ) \in \mathcal{B}_k$, we substitute the pair $(i_\circ, j_\circ)$ by some pair (i', j') according to the rules of the dual method of successive improvement, and thus obtain the set $\mathcal{B}_{k+1}$. We refer to this process as the *adjacent vertex method* and prove that this method converges if the problem (3.1)–(3.4) is dual non-degenerate: for every dual feasible solution, the number of inequalities of the dual problem which turn into equalities at this solution is not greater than $m + n - 1$. This condition assures the uniqueness of the solution of (3.1)–(3.4) for each $p \in \sigma$.

Theorem 4.2 *If the problem (3.1)–(3.4) is dually non-degenerate, then the neighbour vertex method gives an equilibrium point for the cooperation model in a finite number of steps.*

The proof of this statement is based on the fact that during the process, the values of ψ strictly grow. Let us first mention some necessary auxiliary information on the function ψ and the mapping Φ.

Let us say that a set $\mathcal{B} \subset I \times J$ is i-covering if for each $i \in I$ the set $\{j \in J \mid (i,j) \in \mathcal{B}\}$ is non-empty. Denote by $\mathcal{L}$ the collection of dual feasible basis sets of the transportation problem (3.1)–(3.4) and of all their possible

i-covering subsets. Associate to each $\mathcal{B} \in \mathcal{L}$ the set $\Omega(\mathcal{B})$ consisting of all $p \in \sigma$ for which there exists a solution $z_{ij} = z_{ij}^{\mathcal{B}}(p)$ of the system (3.2)–(3.4) satisfying the condition

$$z_{ij} = 0, \quad (i,j) \notin \mathcal{B}. \tag{4.3}$$

All sets $\Omega(\mathcal{B})$, $\mathcal{B} \in \mathcal{L}$, are non-empty. Indeed, $\mathcal{B}_1 \subset \mathcal{B}_2$, $\mathcal{B}_1, \mathcal{B}_2 \in \mathcal{L}$ implies $\Omega(\mathcal{B}_1) \subset \Omega(\mathcal{B}_2)$, so we must just verify that the sets $\Omega(\mathcal{B})$ generated by minimal sets of $\mathcal{L}$ (with respect to the inclusion $\subset$) are non-empty. For a minimal $\mathcal{B} \in \mathcal{L}$, the set I is divided into the subsets $I_j = \{i \in I \mid (i,j) \in \mathcal{B}\}$, and $\Omega(\mathcal{B})$ consists of one point whose coordinates are

$$p_j = \begin{cases} \sum\limits_{i \in I_j} \lambda_i, & \text{if } I_j \neq \emptyset, \\ 0, & \text{if } I_j = \emptyset. \end{cases}$$

It is also clear that $\Omega(\mathcal{B})$, $\mathcal{B} \in \mathcal{L}$, are polyhedral sets, and a point $p \in \Omega(\mathcal{B})$ lies in the relative interior $\Omega^\circ(\mathcal{B})$ if and only if all $z_{ij}^{\mathcal{B}}(p) > 0$, $(i,j) \in \mathcal{B}$. Since the transportation problem (3.1)–(3.4) is dually non-degenerate, it has a unique solution for each $p \in \sigma$, and hence the point $p \in \sigma$ lies in the relative interior of only one of the sets $\Omega(\mathcal{B})$, $\mathcal{B} \in \mathcal{L}$.

To sum up, we can say that the sets $\Omega(\mathcal{B})$, $\mathcal{B} \in \mathcal{L}$ form a polyhedral partition of the simplex σ. The function f is linear on each of $\Omega(\mathcal{B})$, $\mathcal{B} \in \mathcal{L}$. Clearly, for each $p \in \Omega^\circ(\mathcal{B})$ the image $\Phi(p)$ is the same set; we denote it by $\Xi(\mathcal{B})$. Here if $\mathcal{B}$ is a basis set, then $\Omega(\mathcal{B})$ is solid at the affine support of σ, and $\Xi(\mathcal{B})$ is a singleton. It is also clear that $\mathcal{B}_1 \subset \mathcal{B}_2$ implies $\Xi(\mathcal{B}_1) \supset \Xi(\mathcal{B}_2)$. It can be shown also that the sets $\Xi(\mathcal{B})$, $\mathcal{B} \in \mathcal{L}$, form a polyhedral partition of σ°.

For the function $\psi(q) = -f^*(\ln q)$ induced by the cooperation model we need in the proof of the theorem the following representation obtained from the definition of the conjugate function:

$$\psi(q) = \min_{p \in \sigma}\{-(p, \ln q) + f(p)\}, \tag{4.4}$$

and the minimum is attained for all $p \in \Phi^{-1}(q)$. This means that for all $\mathcal{B} \in \mathcal{L}$, the function ψ coincides with $\psi_p(q) = -(p, \ln q) + f(p)$ on $\Xi(\mathcal{B})$ whenever $p \in \Omega(\mathcal{B})$.

Proof of Theorem 4.2. Let $\mathcal{B}_k \in \mathcal{L}$ be a basis set and let $\{p^k\} = \Xi(\mathcal{B}_k)$. In the case $z_{i_\circ j_\circ}^{\mathcal{B}_k}(p^k) < 0$, we use the procedure of the dual method of successive improvement to pass to the basis set

$$\mathcal{B}_{k+1} = \{(i', j')\} \cup \mathcal{B}_k \setminus \{(i_\circ, j_\circ)\}, \quad \mathcal{B}_{k+1} \in \mathcal{L},$$

and to the point p^{k+1} such that $\{p^{k+1}\} = \Xi(\mathcal{B}_{k+1})$. Since the problem is dually non-degenerate, it follows that $p^{k+1} \neq p^k$.

It is clear that the set $\mathcal{B}' = \mathcal{B}_k \setminus \{i_\circ, j_\circ\} = \mathcal{B}_k \cap \mathcal{B}_{k+1}$ also is in $\mathcal{L}$. Consider the set $\Omega' = \Omega(\mathcal{B}') = \Omega(\mathcal{B}_k) \cap \Omega(\mathcal{B}_{k+1})$. It can be easily seen that for $a \in \Omega'$ we have $\ln p^k$, $\ln p^{k+1} \in V(a)$, i. e., p^k, $p^{k+1} \in \Xi(\mathcal{B}')$. But then for the function $\psi_a(q) = (a, \ln q) + f(a)$ we have $\psi(p^k) = \psi_a(p^k)$ and $\psi(p^{k+1}) = \psi_a(p^{k+1})$. Thus, the needed inequality $\psi(p^{k+1}) > \psi(p^k)$ is equivalent to $\psi_a(p^{k+1}) > \psi_a(p^k)$. Let us prove the latter inequality.

As it has been mentioned above, the function f is linear on each of the sets $\Omega(\mathcal{B})$, $\mathcal{B} \in \mathcal{L}$. Since $p^k \in \Xi(\mathcal{B}_k)$ and $p^{k+1} \in \Xi(\mathcal{B}_{k+1})$, we see that $f(p)$ is equal to $(p, \ln p^k) + \Delta_k$ on $\Omega(\mathcal{B}_k)$ and to $(p, \ln p^k) + \Delta_{k+1}$ on $\Omega(\mathcal{B}_{k+1})$; here Δ_k and Δ_{k+1} are some constants. Since f is convex, and $p^k \neq p^{k+1}$, we have

$$\begin{aligned} &(p, \ln p^k - \ln p^{k+1}) > \Delta \quad \text{for } p \in \Omega^\circ(\mathcal{B}_k), \\ &(p, \ln p^k - \ln p^{k+1}) = \Delta \quad \text{for } p \in \Omega(\mathcal{B}_k) \cap \Omega(\mathcal{B}_{k+1}), \end{aligned}$$

where $\Delta = \Delta_k - \Delta_{k+1}$. On the other hand,

$$\begin{aligned} &z^{\mathcal{B}_k}_{i_\circ j_\circ}(p) > 0 \quad \text{for } p \in \Omega^\circ(\mathcal{B}_k), \\ &z^{\mathcal{B}_k}_{i_\circ j_\circ}(p) = 0 \quad \text{for } p \in \Omega(\mathcal{B}_k) \cap \Omega(\mathcal{B}_{k+1}). \end{aligned}$$

Taking into account that the dimension of $\Omega(\mathcal{B}_k) \cap \Omega(\mathcal{B}_{k+1}) = \Omega(\mathcal{B}')$ is $n-2$, we conclude that the conditions

$$z^{\mathcal{B}_k}_{i_\circ j_\circ}(p) < 0$$

and

$$(p, \ln p^k - \ln p^{k+1}) < 0$$

are equivalent on σ. Thus,

$$(p^k, \ln p^k - \ln p^{k+1}) < \Delta.$$

But due to a property of the entropy, $(p^k, \ln p^k) > (p^k, \ln p^{k+1})$ and hence $\Delta > 0$. Now we obtain

$$\psi_a(p^{k+1}) - \psi_a(p^k) = -(a, \ln p^{k+1} - \ln p^k) = \Delta > 0$$

for $a \in \Omega(\mathcal{B}_k) \cap \Omega(\mathcal{B}_{k+1})$, which was to be proved.

Thus, the value of ψ strictly increases when passing from p^k to p^{k+1}. Since these points are uniquely generated by the basis sets $\mathcal{B}_k$ and $\mathcal{B}_{k+1}$, and the number of possible basis sets is finite, so is the process. □

For the exchange model, the functions φ and ψ are strictly convex. Thus, the exchange model has a unique equilibrium point which is the common minimum point of these functions. Thus, to find the equilibrium in the exchange model, we can use any minimization procedure for φ and ψ. In particular, using the way these functions were defined, we can build finite methods based on suboptimization ideas (Rubinstein and Shmyrëv, 1971). Let us describe one of these procedures minimizing the function ψ.

It is not difficult to obtain a more explicit description of the sets $\Xi(\mathcal{B})$ not using the notion of the optimal solution of the transportation problem (3.1)–(3.4). Indeed, it can be easily seen that for $p \in \Omega^\circ(\mathcal{B})$, the set of vectors $w = (u_1, \dots, u_m, v_1, \dots, v_n) \in W(p)$ will be described by the system

$$\begin{aligned} u_i &= \ln c_{ij} - v_j, \quad (i,j) \in \mathcal{B}; \\ u_i &= \operatorname{opt}_{j \in J} (\ln c_{ij} - v_j). \end{aligned}$$

Denoting by y_i the values e^{u_i}, we obtain the following conditions on $q = e^v$ for $v \in V(p)$:

$$\begin{aligned} y_i &= \tfrac{c_{ij}}{q_j}, \qquad\qquad (i,j) \in \mathcal{B}; \\ y_i &= \operatorname{opt}_{j \in J} \tfrac{c_{ij}}{q_j}, \quad i \in I. \end{aligned}$$

Thereby, we obtain the following description of $\Xi(\mathcal{B})$:

$$\Xi(\mathcal{B}) = \left\{ q \in \sigma^\circ \mid \operatorname{opt}_{k \in J} \frac{c_{ik}}{q_k} = \frac{c_{ij}}{q_j}, (i,j) \in \mathcal{B} \right\}.$$

It is important to note that $\Xi(\mathcal{B})$ is the intersection of some polyhedral set and σ°, and lies in the affine manifold $M(\mathcal{B})$ given by the system of linear equations

$$\frac{p_j}{c_{ij}} = \frac{p_l}{c_{il}}, \quad (i,j),\ (i,l) \in \mathcal{B}. \tag{4.5}$$

Let us describe also the affine manifold $L(\mathcal{B})$ whose intersection with σ defines the affine support of the polyhedron $\Omega(\mathcal{B})$. To do it, we introduce the graph $\Gamma(\mathcal{B})$ whose set of vertices is $G = \{1, 2, \dots, m+n\}$ and the set of edges is $\{(i, m+j) \mid (i,j) \in \mathcal{B}\}$. Let τ be the number of components of this graph, G_ν be the set of vertices of the νth component, $I_\nu = I \cap G_\nu$ and $J_\nu = \{j \in J \mid (m+j) \in G_\nu\}$. It is not difficult to show that the following system of linear equations must hold for $p \in \Omega(\mathcal{B})$:

$$\sum_{j \in J_\nu} p_j = \sum_{i \in I_\nu} \lambda_i, \quad \nu = 1, \dots, \tau. \tag{4.6}$$

This system just describes the manifold $L(\mathcal{B})$.

Now let us pass to the description of the algorithm minimizing the function ψ. Let $\mathcal{B}_k \in \mathcal{L}$, and let us know some point $q^k \in \Xi(\mathcal{B}_k)$. We may choose the initial point q^0 as an arbitrary point of σ° and form the initial $\mathcal{B}_0$ as

$$\mathcal{B}_0 = \left\{ (i,j) \in I \times J \mid \max_{k\in J} \frac{c_{ik}}{q_k^0} = \frac{c_{ij}}{q_j^0} \right\}.$$

Given $\mathcal{B}_k$, we find $p = r^k \in \sigma$ satisfying (4.5)–(4.6), i. e., $r^k \in L(\mathcal{B}_k) \cap M(\mathcal{B}_k) \cap \sigma$. As it can be easily seen, the point p is uniquely determined: indeed, on the νth component of the graph $\Gamma(\mathcal{B})$, the values p_j, $j \in I_\nu$, are defined by (4.5) up to a constant multiplier which can be found from the respective equation (4.6). Two cases are possible:

(*i*) $q^k = r^k$;

(*ii*) $q^k \neq r^k$.

In case (*i*), for $p = q_k$ and $\mathcal{B} = \mathcal{B}_k$, the system of conditions (3.2), (3.3), (4.3) turns out to be compatible, and we define $z_{ij} = z_{ij}^{\mathcal{B}_k}(q^k)$. If all z_{ij} are non-negative, then $q^k \in \Omega(\mathcal{B}_k)$, and thus q^k is the equilibrium point of the model. And if $z_{i_\circ j_\circ} < 0$, we put $\mathcal{B}_{k+1} = \mathcal{B}_k \setminus \{(i_\circ j_\circ)\}$, $q^{k+1} = q^k$, and pass to the next step.

In case (*ii*), it may happen that $r^k \in \Xi(\mathcal{B}_k)$. Then we put $q^{k+1} = r^k$, $\mathcal{B}_{k+1} = \mathcal{B}_k$, and pass to case (*i*). If, conversely, $r^k \notin \Xi(\mathcal{B}_k)$, then we put $q(t) = q^k + t(r^k - q^k)$ and find $t^* = \max\{t \mid q(t) \in \Xi(\mathcal{B}_k)\}$. If in this case the condition

$$\frac{q_{j_1}(t)}{c_{i_1 j_1}} \geq \frac{q_l(t)}{c_{i_1 l}}, \quad (i_1, l) \in \mathcal{B}_k,\ (i_1, j_1) \notin \mathcal{B}_k, \tag{4.7}$$

turns out to be limiting, then we put $q^{k+1} = q(t^*)$, $\mathcal{B}_{k+1} = \mathcal{B}_k \cup \{(i_1, j_1)\}$, and pass to the next step.

Theorem 4.3 *If the problem is dually non-degenerate, the described suboptimization method leads to an equilibrium price vector of the exchange model in a finite number of steps.*

Proof. First, let us prove that the function ψ does not grow at the sequence q^k. As it has been mentioned, according to (4.4), the function ψ coincides with all ψ_p on $\Xi(\mathcal{B}_k)$ for $p \in \Omega(\mathcal{B}_k)$. Let us make sure that r^k is the point of the minimum on $M(\mathcal{B}_k) \cap \sigma$ for each of these functions. We may pass from the functions $\psi_p(q)$ to the functions $\bar{\psi}_p(q) = \bar{\psi}_p(q) + \ln \sum_{j\in J} q_j$ which coincide with $\psi_p(q)$ on σ. In their turn, the functions $\bar{\psi}_p$ are positively homogeneous functions of degree 0 , and hence we may consider $I\!R^n_+$ instead of σ.

Let us fix an arbitrary point $p = p^\circ \in \Omega(\mathcal{B}_k)$. Note that if some coordinate p°_j of p° is equal to zero, then the function $\bar{\psi}_{p^\circ}(q)$ is monotone with respect to q_j, and clearly, in the point of the minimum of this function on $M(\mathcal{B}_k) \cap I\!R^n_+$, the coordinate $\bar{q}_j$ is also zero. Restricting $\bar{\psi}_{p^\circ}$ to the subspace $q_j = 0$, we obtain a function of the same type on the space of smaller dimension. Thus, we may consider only the case $p^\circ > 0$, which clearly implies $r^k > 0$.

The set $M(\mathcal{B}_k) \cap I\!R^n_+$ is the conic hull of the points s^ν, $\nu = 1, \ldots, \tau$. Each of them is determined by the respective component of the graph $\Gamma(\mathcal{B}_k)$ described above:

$$s^\nu_j = r^k_j, \quad j \in J_\nu, \tag{4.8}$$

$$s^\nu_j = 0, \quad j \notin J_\nu. \tag{4.9}$$

Now it is clear that r^k is the minimum point of $\bar{\psi}_{p^\circ}$ if and only if the derivative of this function on each direction s^ν, $\nu = 1, \ldots, \tau$, is equal to zero at the point r^k. We have

$$\mathrm{grad}\bar{\psi}_{p^\circ}(r^k) = -\begin{pmatrix} p^\circ_1/r^k_1 \\ \vdots \\ p^\circ_n/r^k_n \end{pmatrix} + \frac{1}{\sum\limits_{j\in J} r^k_j}\begin{pmatrix} 1 \\ \vdots \\ 1 \end{pmatrix}.$$

Since $r^k \in \sigma$, and thus $\sum\limits_{j\in J} r^k_j = 1$, we obtain

$$\frac{\partial\bar{\psi}_{p^\circ}(r^k)}{\partial s^\nu} = \sum_{j\in J_\nu} \frac{p^\circ_j}{r^k_j} r^k_j + \sum_{j\in J_\nu} r^k_j = -\sum_{j\in J_\nu} p^\circ_j + \sum_{j\in J_\nu} r^k_j = 0,$$

because p°, $z^k \in L(\mathcal{B}_k)$.

So, r^k is the minimum point of the functions ψ_p, $p \in \Omega(\mathcal{B}_k)$, on the set $M(\mathcal{B}_k) \cap \sigma$. This implies that when we pass from q^k to q^{k+1}, the values of these functions do not increase. Since q^{k+1}, $q^k \in \Xi(\mathcal{B}_k)$, this is equivalent to the fact that the value of ψ does not increase. Since the functions $\bar{\psi}_p$ are strictly convex on σ, in case $q^{k+1} \neq q^k$ we have $\psi(q^{k+1}) < \psi(q^k)$. Now let us show that if the current point of the process is not yet the minimum point of ψ needed, then the situation of $q^{k+1} \neq q^k$ appears after a finite number of steps.

Since in case *(ii)* the number of elements of $\mathcal{B}_k$ only increases, this case may successively repeat only a finite number of times. Thus, without loss of generality we may assume that case (i) occurs for the current point q^k. If $z_{i_\circ j_\circ} < 0$ in the solution of the system (3.2), (3.3), (4.3), we pass to the next step with $\mathcal{B}_{k+1} = \mathcal{B}_k \setminus \{(i_\circ, j_\circ)\}$ and $q^{k+1} = q^k$. Let us describe the direction from r^k to r^{k+1}.

Consider the graphs $\Gamma(\mathcal{B}_k)$ and $\Gamma(\mathcal{B}_{k+1})$. The graph $\Gamma(\mathcal{B}_{k+1})$ is obtained from $\Gamma(\mathcal{B}_k)$ by deleting the edge $(i_\circ, m+j_\circ)$; the component containing this edge is thus divided to two. Let this component of $\Gamma(\mathcal{B}_k)$ be the first one, containing the vertices $i \in I_1$ and $(m+j)$ for $j \in J_1$. If we exclude the edge $(i_\circ, m+j_\circ)$, the sets I_1 and J_1 are divided to I_1', I_1'' and J_1', J_1'' respectively. Let $i_\circ \in I_1'$ and thus $j_\circ \in J_1''$. Since λ_i is positive, it follows from the definition of r^k that the condition $I_1 \neq \emptyset$ implies $r_j^k > 0$, $j \in J_1$. Let us compute the value $\mu = \sum_{j \in J_1'} r_j^k / \sum_{j \in J_1''} r_j^k$ and construct the vector s whose components are

$$s_j = \begin{cases} -r_j^k, & j \in J_1'; \\ \mu r_j^k, & j \in J_1''; \\ 0, & j \notin J_1. \end{cases}$$

It can be verified easily that this vector defines the direction from r^k to r^{k+1}. Indeed, $\sum_{j=1}^{n} s_j = 0$ and thus the direction s does not lead out from the affine support of the simplex σ. The coordinates q_j of the point q corresponding to the vertices $(m+j)$ of each component of $\Gamma(\mathcal{B}_{k+1})$ change proportionally when we move from q at the direction s. This means that s does not lead out from $M(\mathcal{B}_{k+1})$. Let us verify that for an appropriate $t = \hat{t} > 0$, the point $q(t) = r^k + ts$ belongs to $L(\mathcal{B}_{k+1})$. The following condition is sufficient for that:

$$\sum_{j \in J_1'} q_j(\hat{t}) = \sum_{i \in I_1'} \lambda_i.$$

From this condition we obtain that

$$\hat{t} = \frac{\sum_{j \in J_1'} r_j^k - \sum_{i \in I_1'} \lambda_i}{\sum_{j \in J_1'} r_j^k} \tag{4.10}$$

and so

$$z_{i_\circ j_\circ} = \sum_{i \in I_1'} \lambda_i - \sum_{j \in J_1'} r_j^k < 0$$

implies $\hat{t} > 0$. Thus, the direction s obtained leads from the point $q^{k+1} = r^k$ to the point r^{k+1}. Next, let us consider the possibility of a positive shift at this direction within the set $\Xi(\mathcal{B}_{k+1})$.

Each pair $(i,j) \notin \mathcal{B}_{k+1}$ induces a constraint on the size of the shift:

$$\frac{c_{ij}}{q_j(t)} \leq \frac{c_{il}}{q_l(t)}, \quad (i,l) \in \mathcal{B}_{k+1}. \tag{4.11}$$

Clearly, the constraint induced by $(i_\circ, j_\circ)$ is not limiting: the value $q_{j_\circ}(t)$ increases when t grows since $j_\circ \in J_1''$, and the value $q_l(t)$ decreases for $(i_\circ, l) \in \mathcal{B}_{k+1}$ since $l \in J_1'$.

Let the limiting constraint be the inequality (4.11) induced by the pair $(i,j) = (i_1, j_1)$. We shall pass to the next step setting $\mathcal{B}_{k+2} = \mathcal{B}_{k+1} \cup \{(i_1, j_1)\}$. If the shift turned out to be equal to zero, i. e., if $q^{k+2} = q^{k+1}$, this means that

$$\frac{c_{i_1 j_1}}{q_{j_1}^{k+1}} = \frac{c_{il}}{q_l^{k+1}}, \quad (i,l) \in \mathcal{B}_{k+1}, \tag{4.12}$$

and thus the point $q^{k+1} (= q^k = r^k)$ is in the set $\Xi(\bar{\mathcal{B}}_k)$ for $\bar{\mathcal{B}}_k = \mathcal{B}_k \cup \{(i_1, j_1)\}$. At the same time, $q^{k+1} \in L(\mathcal{B}_k) \subset L(\bar{\mathcal{B}}_k)$. So, if we start a step of the process with the set $\mathcal{B}_k \in \mathcal{L}$, and the point $\bar{q}^k = q^{k+1} \in \Xi(\bar{\mathcal{B}}_k)$, we obtain case (i). From this, solving the system (3.2), (3.3), (4.3) for $p = \bar{q}^k$ and $\mathcal{B} = \bar{\mathcal{B}}_k$, we obtain the values $\bar{z}_{ij} = z_{ij}^{\bar{\mathcal{B}}_k}(\bar{q}^k)$ coinciding with the values $z_{ij}^{\mathcal{B}_k}(q^k)$. So, we still have $\bar{z}_{i_\circ j_\circ} < 0$ and pass to the next step with the set $\bar{\mathcal{B}}_{k+1} = \bar{\mathcal{B}}_k \setminus \{(i_\circ j_\circ)\} = \mathcal{B}_{k+2}$ and $\bar{q}^{k+1} = q^{k+2} (= q^{k+1})$. Thus, the situation at step $(k+2)$ is qualitatively the same as it was at step $(k+1)$, and hence the constraint on the size of the shift induced by $(i_\circ, j_\circ)$ is also not limiting.

It is clear that steps of the form described, i. e., having zero size of the shift, may repeat only a finite number of times. Indeed, the current set $\mathcal{B}_s$ grows after such a step, and since the problem (3.1)–(3.4) is dually non-degenerate, if $\mathcal{B}_k$ is a basis set, then there are no pairs (i_1, j_1) for $\mathcal{B}_{k+1} = \mathcal{B}_k \setminus \{(i_\circ, j_\circ)\}$ for which (4.12) would hold.

We have shown that the value of ψ in the current point of the process strictly decreases between two successive occurrences of case (i). However, in case (i) this value is uniquely determined by the set $\mathcal{B}_k$. So, no one of the sets occurring in a realization of case (i) can occur in the following once more. Therefore, since the number of possible sets $\mathcal{B} \in \mathcal{L}$ is finite, so is the number of steps in the process. $\square$

5 Some generalizations

Since the function f generated by the cooperation model is convex, it follows that for all p^1, $p^2 \in \sigma$ and $q^1 \in \Phi(p^1)$, $q^2 \in \Phi(p^2)$ the inequality holds

$$(p^1 - p^2, \ln q^1 - \ln q^2) \geq 0.$$

It can be written concisely as

$$(p^1 - p^2, \ln \Phi(p^1) - \ln \Phi(p^2)) \geq 0.$$

For the exchange model, the analogous inequality of the opposite sign holds. Just this property of the mapping Φ, which can be conventionally called logarithmic monotonicity, provides the basis for the constructions above allowing to find the fixed points of such mappings. The potential property used, due to which there exists a function $f : \mathbb{R}^n \to \mathbb{R}^1$ such that $\partial f(p) = \ln \Phi(p) + \lambda\theta$, $\lambda \in \mathbb{R}^1$, for $p \in \sigma$, follows from the fact that the mapping Φ is logarithmically monotone. So, the considerations above can be generalized to arbitrary logarithmically monotone piecewise constant mappings of the simplex σ to itself. The mappings generated by the exchange or cooperation model are just particular cases of this situation. Let us describe more precisely the class of mappings we mean.

As was mentioned in Shmyrëv (1981b), we may introduce a linear space structure on σ° by defining the sum of the elements p^1 and p^2 of σ° as the point p whose coordinates are $p_j = p_j^1 p_j^2 / \sum_{k=1}^{n} p_k^1 p_k^2$, and the product of $p \in \sigma^\circ$ and a real scalar t as the point q whose coordinates are $q_j = p_j^t / \sum_{k=1}^{n} p_k^t$. Denote this linear space by Σ. Thus, we can consider not only usual polyhedral sets of the space $\mathbb{R}^n$ in σ°, but also polyhedral sets of the space Σ, i. e., the sets of solutions of finite systems of inequalities linear on Σ. Linear functionals in Σ are of the form $\alpha(p) = \sum_{i=1}^{n} \alpha_i \ln p_i$ for $\sum_{i=1}^{n} \alpha_i = 0$.

Let us consider a polyhedral complex ξ on Σ whose cells constitute a partition of Σ , i. e., each point $q \in \Sigma$ belongs to the relative interior of exactly one cell of ξ. Consider a point-to-set mapping $F : \sigma^\circ \to 2^\sigma$ constant on the relative interior Ξ° of each cell $\Xi \in \xi$ and having the following properties:

(a) for full-dimensional cells $\Xi \in \xi$, the image $F(\Xi^\circ)$ is a singleton;

(b) for each cell $Q \in \xi$, the equality holds

$$F(Q^\circ) = \text{conv}\{F(\Xi^\circ) \mid \Xi \supset Q, \Xi \in \xi\},$$

where the convex hull is taken in the sense of $I\!R^n$.

Let us say that F is a piecewise constant logarithmically monotone increasing mapping if the following condition holds:

(c) for all points $p, q \in \sigma^\circ$,

$$(\ln p - \ln q, v - w) \geq 0 \quad \forall v \in F(p),\ \forall w \in F(q),$$

or briefly

$$(\ln p - \ln q, F(p) - F(q)) \geq 0.$$

Converting the sign of the inequality (c) to the opposite, we obtain another class of mappings which we call decreasing logarithmically monotone piecewise constant mappings.

It can be easily seen that the mapping $F = \Phi^{-1}$ is of the first or the second class, depending on which of the problems is considered, the cooperation or the exchange one. Such mappings have two special features:

1. the polyhedrons $\Omega = F(Q^\circ)$, $Q \in \xi$, cover the whole simplex σ and do not intersect at relatively interior points;

2. linear functionals of the space Σ specifying the polyhedral sets $\Xi \in \xi$ have the special form $\alpha(p) = \ln p_i - \ln p_j$, so these sets can be given by systems of inequalities linear in $I\!R^n$.

A study of piecewise constant logarithmically monotone mappings on a simplex is reduced to a study of monotone piecewise constant mappings in $I\!R^n$. As it was proved by the author in Shmyrëv (1981c), the latter mappings are potential. Using this fact, one can generalize the described procedures for searching fixed points. For example, if F is an increasing logarithmically monotone mapping, and $F(\sigma^\circ) \subset \sigma^\circ$, then the fixed point of this mapping can be obtained by the iteration method $p^{k+1} \in F(p^k)$.

Acknowledgement This is a slightly adapted translation from Russian of the paper 'Algorithms of searching for the equilibrium in fixed budget exchange models' originally published in Optimizacija 31, 1983, 137–155.

References

1. Eaves, B.C. (1976): "A finite algorithm for the linear exchange model," *Journal of Mathematical Economics*, 3, 197–204.

2. Gale, D. (1960): *The Theory of Linear Economic Models.* McGrawhill Company, Inc., New York-Toronto-London.

3. Lemke, C.E. (1965): "Bimatrix equilibrium points and math. programming," *Management Science*, 7.

4. Negishi, T. (1960): "Welfare economics and existence of an equilibrium for a competitive economy," *Metroeconomica*, 12, 92–97.

5. Rockafellar, R.T. (1970): *Convex analysis.* Princeton Univ. Press, Princeton, New Jersey.

6. Rubinstein, G.Sh., and V.I. Shmyrëv (1971): "Methods of minimization of a quasiconvex function on a convex polyhedron," *Optimizatsia*, 1, 82–117 (in Russian).

7. Shmyrëv, V.I. (1981a): "On the determination of fixed points of piecewise constant monotone mapping in $\mathbb{R}^n$," *Soviet Math. Dokl.*, 24.

8. Shmyrëv, V.I. (1981b): "Monotonicity in linear exchange models," *Optimizatsia*, 27, 77–95 (in Russian).

9. Shmyrëv, V.I. (1981c): "On the property of piecewise monotone mappings in $\mathbb{R}^n$ to be potential," *Optimizatsia*, 27, 65–76 (in Russian).

10. Shmyrëv, V.I. (1982): "On algorithms for finding fixed points of piecewise constant monotone mappings in $\mathbb{R}^n$," *Optimizatsia*, 29, 32–44 (in Russian).

11. Shmyrëv, V.I. (1983): "On approach to the determination of equilibrium in elementary exchange models," *Soviet Math. Dokl.*, 27, No 1.

Chapter 13

Equilibrated states and theorems on the core

V.I. Danilov and A.I. Sotskov

Abstract: This paper deals with analogs of equilibrated states introduced in Polterovich (1984). We obtain some generalizations and propose a uniform approach to the existence of equilibrated states and the core of cooperative games.

Key words: equilibrium, core, effectivity function.

1 Introduction

The notion of equilibrated states was introduced in Polterovich (1984). We consider here some of its analogs and generalizations. To explain the situations in which these analogs may emerge, let N be a set of agents and X be a set of alternatives. Suppose each agent has a preference relation on X. We are interested in which element of X should be chosen? If the only available information is the agents' preferences, then the answer would be to choose an arbitrary Pareto-optimal alternative (observe that such an alternative exists under very weak assumptions). However, typically there are many Pareto efficient alternatives, so that it should be useful to have a more refined choice. In many cases the considered situation contains additional information that puts the agents in different positions depending on the alternatives. For example, in a resource allocation problem, the additional information may contain the prices of resources, the incomes of the agents or their initial endowments, see Polterovich (1984).

Based on such additional information it maybe possible, given alternative x, to distinguish a certain coalition of agents, which has a so-called "decisive word" with respect to alternative x. For instance, in the resource allocation problem with given incomes, the decisive word with respect to an allocation x can be given to a coalition of agents getting the bundles with values less than their incomes. Also, when the initial endowments are known, the decisive word can be given to a coalition which can improve on x by reallocating the total initial endowment of the members of the coalition. Now, an alternative is said to be "equilibrated" when it is not rejected by the coalition having the decisive word for this alternative.

Within such a framework, it looks as if alternatives have their "favorites" among the agents. That is not only the agents have preferences and choose alternatives, but also the alternatives in their turn prefer and choose agents. The symmetry becomes even stronger when a singleton equilibrated state does not exist, and we have to proceed to subsets of alternatives. In this case a coalition $K \subset N$ of agents and a bunch $Y \subset X$ of alternatives are considered as equilibrated when their choices are reciprocally coordinated, i.e., K chooses Y and Y chooses K.

The paper is organized as follows. Section 3 is devoted to the formal exposition of the above considerations. Based on general results about maximal elements obtained in Section 2, in Section 3 we give conditions guaranteeing the existence of equilibrated states. The final Section 4 contains some applications to cooperative games.

2 Maximal elements

Let X be a set and let $P \subset X \times X$ be a binary relation on X. It is convenient to represent P as a correspondence from the set X into itself, $P : X \Longrightarrow X$. In particular, we write

$$P(x) = \{x' \in X,\ (x, x') \in P\}, \quad P^{-1}(x) = \{x' \in X,\ (x', x) \in P\}$$

for $x \in X$,

$$P(A) = \bigcup_{x \in A} P(x)$$

for $A \subset X$, and so on.

It is supposed that P is a strict preference relation, so $(x, x') \in P$, xPx', $x' \in P(x)$ (or $x \prec x'$) are all different notations with the meaning that x' is strictly preferred to x. Given a preference P, an element $x^* \in X$ is called *maximal* if $P(x^*) = \emptyset$, that is there are no elements in X better than x^*.

We begin with an elementary and well-known statement about the existence of maximal elements.

Proposition 2.1 *Let X be a finite set and let the relation P on X be acyclic. Then there exists a maximal element in X.*

Remark 2.2 The assertion can be easily reformulated for compact sets, one should only require the relation P to be lower semi-open. (A correspondence $P : X \Longrightarrow Y$ between topological spaces X and Y is *lower semi-open* (l.s.o.) if the set $P^{-1}(y)$ is open for any $y \in Y$.)

Remark 2.3 The acyclicity condition of the relation P can be rewritten in the following way. Let P^∞ denote the transitive closure of P. Then acyclicity of P is equivalent to the condition $x \notin P^\infty(x)$ for any x and can be understood as some "generalized" irreflexivity. The next proposition represents a convex variant of the previous one, and the generalized irreflexivity has the form $x \notin \mathrm{co}P(x)$, where "co" means the convex hull.

Proposition 2.4 *Let X be a convex compact set and let P be a l.s.o. relation on X. Suppose that $x \notin coP(x)$ for any $x \in X$. Then there exists a maximal element in X.*

Hereafter X is supposed to be a convex subset in a finite dimensional vector space, although many assertions remain true in a more general case. Proposition 2 in the given form belongs to Bergstrom (see Kiruta *et al.*, 1980), an equivalent result in another form was obtained by Ky Fan. Below we give a more general assertion.

Let B and P be two relations on the set X, and $P \subset B$. We say that $x^* \in X$ is an *equilibrium* if $x^* \in B(x^*)$ and $P(x^*) = \emptyset$.

Theorem 2.5 *Let X be a convex compact set, and let $P \subset B$ be two relations on X satisfying*

(i) *the relation B is closed and the sets $B(x)$ are convex and nonempty;*

(ii) *the relation P is l.s.o. and $x \notin \mathrm{co}P(x)$.*

Then there exists an equilibrium $x^* \in X$.

Particular cases of the theorem are the Bergstrom-Ky-Fan theorem ($B = X \times X$) and the Kakutani theorem ($P = \emptyset$). The proof is based on Kakutani's theorem.

Proof. Let $F = \{x \in X,\ x \in B(x)\}$ be the set of fixed points of B. It is clear that F is closed in X and is non-empty. Suppose that the theorem is not true. Then, for any point $x \in F$, there exists $y \in P(x)$, or equivalently $x \in P^{-1}(y)$. The latter means that the open (due to l.s.o. of P) sets $P^{-1}(y)$ cover F. Since F is compact, one can choose a finite subcovering $P^{-1}(y_1), \ldots, P^{-1}(y_n)$. If we define $U = \bigcup_{i=1}^{n} P^{-1}(y_i)$, then the set U is open and contains F.

For each $i = 1, ..., n$, let s_i be continuous functions on X equal to zero out of $P^{-1}(y_i)$ and strictly positive on $P^{-1}(y_i)$. Define the mapping $s : U \to X$ by

$$s(x) = (\sum_i s_i(x) y_i)/(\sum_i s_i(x)).$$

Obviously the mapping s is continuous and $s(x) \in \mathrm{co}P(x) \subset B(x)$ for $x \in U$. Now we form the correspondence $\widetilde{B}(x) : X \Longrightarrow X$ by

$$\widetilde{B}(x) = \begin{cases} s(x), & \text{if } x \in U, \\ B(x), & \text{if } x \notin U. \end{cases}$$

It is easy to see that the correspondence $\widetilde{B}$ is closed and has nonempty convex values. So by Kakutani's theorem there exists a fixed point $x^* \in \widetilde{B}(x^*)$. If $x^* \notin U$, then $x^* \in B(x^*)$, that is $x^* \in F$, which contradicts the inclusion $F \subset U$. If $x^* \in U$, then $x^* \in \mathrm{co}P(x^*)$, which contradicts condition (ii). The obtained contradiction proves the theorem. □

Remark 2.6 Below we are going to reduce finite problems to convex ones. As a sample of such reduction we show that Proposition 2.1 can be reduced to Proposition 2.4.

Given a finite set X, denote by ΔX the simplex generated by X, that is the subset in $\mathbb{R}^X$ consisting of vectors $\lambda = (\lambda_x)_{x \in X}$, where $\lambda_x \geq 0$ and $\sum_{x \in X} \lambda_x = 1$. Elements of X are naturally identified with vertices of the simplex ΔX. Any binary relation P on X can be naturally extended on ΔX by the formula:

$$\tilde{P}(\lambda) = P(\mathrm{supp}(\lambda)),$$

where $\mathrm{supp}(\lambda) = \{x \in X, \lambda_x > 0\}$ is the support of λ. As it is easy to see, the relation $\tilde{P}$ is l.s.o. on ΔX. Moreover, $\lambda \notin \mathrm{co}\tilde{P}(\lambda)$. To show this, suppose that λ is a convex combination of elements $x_1, \ldots, x_n \in \tilde{P}(\lambda)$. Then $Y = \mathrm{supp}(\lambda)$ is contained in $\{x_1, \ldots, x_n\}$ so that $Y \subset P(Y)$. This contradicts the acyclicity of the relation P.

Thus, the relation $\tilde{P}$ on ΔX satisfies the assumptions of Proposition 2.4 and hence there exists a maximal element $\lambda^* \in \Delta X$. Therefore any x

from the support of λ is maximal element of X with respect to P, that is $P(x) = \emptyset$. So, Proposition 2.1 holds.

In the above consideration, convex combinations of elements of X act through their supports, that is through subsets of X. We would like to develop this idea further. Let $\hat{X}$ denote the set of non-empty subsets of X, $\hat{X} = 2^X \setminus \{\emptyset\}$; elements of $\hat{X}$ are called *bunches*. In a sense, the set $\hat{X}$ is an analog of simplex ΔX. Again X is canonically embedded into $\hat{X}$. The elements of $\hat{X}$ play (in a rough form) the role of "intermediate" or compromise alternatives. Again any binary relation P on X can be naturally extended to the following relation $\hat{P}$ on $\hat{X}$:

$$(A, B) \in \hat{P} \Leftrightarrow B \subset P(A) \Leftrightarrow \forall\, b \in B\ \exists\, a \in A \text{ such that } (a, b) \in P.$$

The relation $\hat{P}$ is monotone with respect to the natural order structure on $\hat{X}$ (inclusion), that is $A \subset B$ implies $\hat{P}(A) \subset \hat{P}(B)$. This property motivates the following proposition.

Proposition 2.7 *Let $(X, \leq)$ be a non-empty finite partially ordered set containing the least upper bound* (sup) *of any non-empty subset of X. Let P be a binary relation on X satisfying:*
(i) if $x \leq y$ then $P(x) \subset P(y)$;
(ii) $x \not\leq \sup P(x)$ for any $x \in X$.
Then there exists a maximal element in X with respect to P.

Proof. The proof goes along the lines of Birkhoff-Tarski's theorem. Suppose the assertion is not true, that is $P(x)$ is not empty for any $x \in X$, and define $p(x) = \sup P(x)$. Then p is monotone, that is $x \leq y$ implies $p(x) \leq p(y)$. Let $e = \sup X$ be the greatest element of X. Since $e \geq p(e)$, we get by monotonicity of p the infinite decreasing chain:

$$e \geq p(e) \geq p^2(e) \geq \ldots$$

Here all inequalities are strong because of condition (ii). Therefore all elements of this chain are different. However this contradicts the assumption of finiteness of X, which proves the proposition. □

Note that the monotonicity condition (i) is an analog of the continuity property or l.s.o. The condition (ii) is a kind of "generalized" irreflectivity.[1]

---

[1] This note was developed in our paper Danilov and Sotskov (1998).

3 Equilibrated states

We now proceed to a more complex situation. Let a family of non-empty sets $\{X_i, i \in I\}$ be given. From each set X_i an element x_i should be chosen, where the choice of x_i depends on the choices of other x_j, $j \neq i$. This is a typical problem in game theory, where the sets X_i are interpreted as the sets of strategies.

For sake of simplicity, later on we consider the case of two sets $X_1 = N$ and $X_2 = X$. The elements of N are interpreted as participants and those of X as alternatives. According to Section 1 we are interested in the choice of the most preferable alternatives. We assume that every participant $i \in N$ has a preference relation P_i on the set X. Further, we assume that with every alternative $x \in X$ an exogeneously given set $S(x) \subset N$ of its "favorite" participants is associated. Then preferences of a coalition $S \subset N$ is formed from the individual preferences of its members by the unanimity rule: $P_S = \cap_{i \in S} P_i$. So, according to the unanimity rule, an element $x \in X$ is maximal with respect to P_S when there does not exist another element in X that is uniformly preferred above x by all members of S. Although it may happen that any member of S prefers some other element in X above x, they can not agree on some other element that is better for all members. Observe that this requires less than the Pareto rule, which searches for an alternative such that no member of S can made off better without hurting any of the others. However, when each preference P_i is strict and complete (in the sense that for any two different elements x and y either $(x, y) \in P_i$ or $(x, y) \in P_i$), then the two concepts coincide.

Definition 3.1 A pair (S, x), where $S \subset N$ is a non-empty coalition of participants and $x \in X$ is an alternative, is called an *equilibrated state* if $S = S(x)$ and $P_S(x) = \emptyset$.

In other words, an alternative x is equilibrated if it is the best for its "favorite" coalition $S(x)$, i.e., if an alternative x is approved by its "favorite" coalition. In terms of the introduction we say that the "favorite" coalition has the "decisive word". In the sequel we shortly speak about the "decisive" coalition.

Remark 3.2 Essentially the concept of equilibrated states was introduced by Polterovich (1984). However we would like to note three differing nuances. 1) The coalition preference relations are formed by the unanimity rule and not by the more delicate Pareto rule. 2) Unlike here, Polterovich forms the "decisive" coalitions $S(x)$ as the set of maxima of some functions $g_x(\cdot)$ on

N. Such way of coalition forming has its own advantages (and we come back to it later on). 3) Unlike our paper, Polterovich imposes on an equilibrated state x the additional requirement $S(x) = N$.

Even under natural assumptions like finiteness of the sets N and X, acyclicity of preferences P_i and so on, equilibrated states may fail to exist.

Example 3.3 Let $N = \{1, 2\}$, $X = \{x, y\}$, $P_1 = \{(x, y)\}$, $P_2 = \{(y, x)\}$, $S(x) = \{1\}$ and $S(y) = \{2\}$. Here the state x is not equilibrated since its "decisive" agent 1 prefers y and symmetrically, the state y is not equilibrated.

The absence of equilibrium in this example is caused by the sharp transition from x to y. As we shall see, a presence of compromise variants between x and y (in the form of convex combinations or of bunches) just provides the existence of equilibrated states. We begin with the convex case (cf. Polterovich, 1984).

Proposition 3.4 *Let N be a finite set, and let X be a convex compact set. Assume that*

(i) the correspondence $S : X \Longrightarrow N$ is closed;

(ii) every correspondence P_i is l.s.o. and $x \notin \mathrm{co}P_i(x)$.

Then there exists an equilibrated state.

Proof. Define the correspondence $P : X \Longrightarrow X$ by

$$P(x) = \bigcap_{i \in S(x)} P_i(x).$$

We assert that P satisfies the conditions of Proposition 2.4. The property $x \notin \mathrm{co}P(x)$ is obvious. Let us check that P is l.s.o. Let y belong to $P(x_0)$; we have to show that $y \in P(x)$ for all near points x, i.e. for all points x in a small enough neighborhood of x_0. Let $S_0 = S(x_0)$ and $i \in S_0$; then $y \in P_i(x_0)$. Since P_i is l.s.o. $y \in P_i(x)$ for near points x. The finiteness of N and closedness of the correspondence S imply that $S(x) \subset S_0$ for near x. So, for such a point x and for each $i \in S(x)$, there holds $y \in P_i(x)$, that is $y \in P(x)$. Now Proposition 2.4 implies the existence of an alternative x^* such that $P(x^*) = \emptyset$, which proves the proposition. □

Let us now discuss the case when X is finite and equilibrated states are sought in the form of bunches. In this case we enlarge the initial set X up to the set $\hat{X} = 2^X \setminus \{\emptyset\}$. Preferences P_i are naturally extended on

$\hat{X}$, see Section 2. Besides that we should extend the correspondence of "decisive" coalitions on $\hat{X}$. However there is no indisputable candidate for such extension.

One could act straightforwardly and set for $A \in \hat{X}$

$$\hat{S}(A) = S(A) = \bigcup_{x \in A} S(x).$$

In this case, as we shall see, equilibrated bunches always exist. However, to form a "decisive" coalition of a bunch $A \subset X$ as the union of the "decisive" coalitions $S(x)$, $x \in A$, is not very interesting. In order to allow for more interesting opportunities, one should use more refined ways of evaluating participants by alternatives and also the ways of their aggregation as it was noted in Remark 3.2.

Suppose that the evaluation of the agents of an alternative x is given in the form of a cardinal utility function $g_x : N \to I\!R$ (or in the form of a vector $g_x \in I\!R^N$). The evaluation of a bunch A is given by a convex combination of the utilities of its members: $\sum_{x \in A} \lambda_x g_x$. The coefficients λ_x, that is the weights of the members of the bunch, are not fixed in advance, but represent one of the components of an equilibrated state.

Definition 3.5 A triple (S, Y, g), where $S \subset N$ is a non-empty coalition, $Y \subset X$ is a bunch of alternatives and $g \in \mathrm{co}\{g_x, x \in Y\}$, is called an *equilibrated state* (in bunches) if $S = \mathrm{Argmax}(g)$ and $\hat{P}_S(Y) = \emptyset$.

The latter condition $\hat{P}_S(Y) = \emptyset$ can be rewritten as

$$\hat{P}_S(Y) = \cap_{i \in S}(\cup_{y \in Y} P_i(y)) = \emptyset.$$

Roughly speaking, a triple (S, Y, g) is equilibrated if the coalition S chooses a bunch Y while the bunch Y chooses the coalition S. More precisely, the bunch Y is an S-efficient element of $\hat{X}$, whereas the coalition S consists of the best participants according to the criterion $g : N \to I\!R$ (that is $S = \mathrm{Argmax}(g)$), where g is a mixture of the "utility functions" g_y, $y \in Y$.

Proposition 3.6 *Let N and X be finite sets, $g : X \to I\!R^N$ be a function, and let the preferences P_i be acyclic. Then there exists an equilibrated state in bunches.*

Proof. We shall use the arguments from Remark 2.6. Let ΔX be the simplex spanned on X. The function $g : X \to I\!R^N$ is extended by linearity on ΔX, that is $g(\lambda) = \sum_{x \in X} \lambda_x \, g_x$. The preference relations P_i also are extended on ΔX in the following way: for any $\lambda \in \Delta X$ we set

$$\tilde{P}_i(\lambda) = P_i(\text{supp}(\lambda)).$$

We already checked that $\tilde{P}_i$ is l.s.o. and $\lambda \notin \text{co}\tilde{P}_i(\lambda)$. Define the correspondence $S : \Delta X \Longrightarrow N$ by

$$S(\lambda) = \text{Argmax}(g(\lambda)).$$

It is clear that S is closed. Using Proposition 3.4 we get an equilibrated state (in mixtures) $\lambda^* \in \Delta X$ for which $\tilde{P}_S(\lambda^*) = \emptyset$, where $S = \text{Argmax}(g(\lambda^*))$ and $g(\lambda^*) = \sum \lambda^*_x g_x$. If we set $Y = \text{supp}(\lambda^*)$ and $g^* = g(\lambda^*)$ then the previous relations mean that the triple (S, Y, g^*) is an equilibrated state. □

Remark 3.7 Given the correspondence of "decisive" coalitions $S : X \Rightarrow N$, one can form the functions g_x as the characteristic functions of subsets $S(x) \subset N$ (that is equals 1 on $S(x)$ and 0 outside). Obviously for any $g \in \text{co}(g_y,\ y \in Y)$ it holds that $\text{Argmax}(g) \subset S(Y) = \cup_{y \in Y} S(y)$. So Proposition 3.6 implies the existence of an equilibrated state in the following form: there exists a bunch $Y \subset X$ such that $P_{S(Y)}(Y) = \emptyset$.

Remark 3.8 If the preferences of participants P_i are also given in the form of utility functions $u_i : X \longrightarrow I\!R$ then we get the assertion essentially coinciding with the Scarf lemma, Scarf (1967): *there exist non-empty sets $S \subset N$ and $Y \subset X$ such that:*

a) $S = \text{Argmax}(g)$, where $g \in co(g_y, y \in Y)$;

b) for any $x \in X$ there exists $i \in S$ such that $u_i(x) \leq \min_{y \in Y}(u_i(y))$.

The assumption about finiteness of X can be replaced by that of compactness of X if the functions g_x depend continuously on x. Moreover, a correspondence $G : X \Longrightarrow I\!R^N$ can be introduced instead of functions $g : X \to I\!R^N$.

Theorem 3.9 *Let N be a finite set, X a compact set and $G : X \Longrightarrow I\!R^N$ a correspondence. Assume that*

(i) *$G : X \Longrightarrow I\!R^N$ has a compact graph and non-empty values $G(x)$;*

(ii) *the preferences P_i are l.s.o. and acyclic.*

Then there exists a non-empty $S \subset N$, a non-empty closed $Y \subset X$, and $g \in \text{co}G(Y)$ such that $S = \text{Argmax}(g)$ and $P_S(Y) = \emptyset$.

Proof. Denote by $\hat{X}$ the set of all non-empty closed subsets of X. The set $\hat{X}$ provided with the Hausdorff topology is a compact set. Define the correspondence $\hat{G} : \hat{X} \Longrightarrow I\!R^N$ setting $\hat{G}(Y) = \mathrm{co}(G(Y))$. It is clear that $\hat{G}$ has also a compact graph.

For every $x \in X$ we denote by U_x the set of those pairs $(Y, g) \in \hat{X} \times I\!R^N$ for which $x \in P_{\mathrm{Argmax}(g)}(Y)$. Similarly to the proof of Proposition 3.4, one can check that the sets U_x are open. It is sufficient to show that the inclusion $x \in P_i(Y)$ implies that $x \in P_i(Y')$ for Y' close to Y in Hausdorff topology. However $x \in P_i(Y)$ means that Y intersects the open set $P_i^{-1}(x)$. Then a close to Y set Y' also intersects $P_i^{-1}(x)$.

Thus, the sets U_x are open in $\hat{X} \times I\!R^N$. Suppose that the theorem is false and, for any pair $(Y, g) \in \hat{G}$, the set $P_{\mathrm{Argmax}(g)}(Y)$ is non-empty. Then the family of open sets $\{U_x, x \in X\}$ covers the graph of $\hat{G}$. Due to compactness of $\hat{G}$, one can choose a finite subcovering. In other words, there exists a finite subset $X^0 \subset X$ such that for any pair $(Y, g) \in \hat{G}$ the set X^0 intersects $P_{\mathrm{Argmax}(g)}(Y)$. However this contradicts Proposition 3.6 applied to the set X^0. Precisely, let P_i^0 be the preference P_i constrained on X^0, and $g_x^0 \in G(x)$ for $x \in X^0$. Then according to Proposition 3.6 there exists a non-empty set $Y^0 \subset X^0$ and $g \in \mathrm{co}(g_y^0, y \in Y^0) \subset \hat{G}(Y^0)$ such that the set $P_{\mathrm{Argmax}(g)}(Y^0) \cap X^0$ is empty. The contradiction proves the theorem. □

4 Application to the core

Scarf's lemma and his concept of a primitive set have been a decisive factor at forming our interpretation of equilibrated states. Scarf used his lemma for proving the famous theorem about the non-emptiness of the core of an NTU-game. We also address this issue. As we are interested in coalition aspects of a game, it is important to know which outcomes can be enforced by coalitions. Such information can be presented by effectivity functions.

Let N be a finite set of agents and let X be a set of alternatives. Coalition powers are presented by a correspondence

$$F : 2^N \Longrightarrow 2^X.$$

Informally, a relation $Y \in F(S)$ means that the coalition S can drive the outcome into the set $Y \subset X$. It is naturally to suppose that $\emptyset \notin F(S)$, that $X \in F(S)$, and that F is monotone (that is the inclusions $Y \in F(S)$, $Y \subset Y'$, and $S \subset S'$ imply $Y' \in F(S')$). A correspondence F satisfying these properties is called an *effectivity function.*

Remark 4.1 In order to assimilate the notion of effectivity functions, let us consider an ordinary NTU-game. Such a game is given by a family $\{V_S, S \subset N\}$, where $V_S \subset I\!R^S$ are the sets of "payoff vectors" of a coalition S. We associate with this game the following effectivity function on $X = I\!R^N$: a coalition S enforces each subset $Y \subset X$ containing the set $\pi_S^{-1}(y)$, where $y \in V_S$. Here π_S is the natural projection $I\!R^N \to I\!R^S$.

Other examples of effectivity functions are given by the so called simple games. A simple game is defined by a set of "winning coalitions" $W \subset 2^N$. Assume that coalitions from W can enforce any non-empty subset of X while the other coalitions only can enforce X. This is again an effectivity function.

Given agents' preferences P_i, $i \in N$, on X, one can speak about the core. The core $C(F, P_N)$ consists of all alternatives $x \in X$ such that

$$P_S(x) := \bigcap_{i \in S} P_i(x) \notin F(S)$$

for any non-empty coalition $S \subset N$. In other words, no coalition can improve on a core outcome for all its members. In case of a NTU-game, we come to the ordinary notion of the core.

Now we consider the conditions guaranteeing the core to be non-empty. Let us begin with the balancedness condition analogous to that of Bondareva-Scarf. We recall that a set of coalitions $\mathcal{K} \subset 2^N$ is called *balanced*, if there exists a function $\lambda : \mathcal{K} \longrightarrow I\!R_+$ such that $\sum_{S, i \in S} \lambda(S) = 1$ for any $i \in N$. It is convenient to reformulate this condition as follows: a family $\mathcal{K}$ is balanced if and only if $\delta_N \in \mathrm{co}\{\delta_S, S \in \mathcal{K}\}$, where δ_S is the gravity center of the face ΔS of the simplex $\Delta N \subset I\!R^N$.

Definition 4.2 An effectivity function F is called *balanced* if, for any balanced set of coalitions $\mathcal{K} \subset 2^N$ and any family $\{A_i,\ i \in N\}$ of subsets of X, the following implication holds:

$$\bigcap_{i \in S} A_i \in F(S) \text{ for any } S \in \mathcal{K} \text{ implies } \bigcap_{i \in N} A_i \neq \emptyset.$$

Somewhat loosely speaking, one can understand the balancedness as follows: if the goals of the agents (that is the sets A_i) are attainable by any coalition from a balanced bundle $\mathcal{K}$, then they are not contradictory (that is the sets A_i have a common point).

Theorem 4.3 *Let X be a finite set. If F is a balanced effectivity function, and preferences P_i are acyclic, then the core $C(F, P_N)$ is non-empty.*

Proof. Denote by $H(x)$ (where $x \in X$) the set of non-empty coalitions $S \subset N$ for which

$$P_S(x) = \bigcap_{i \in S} P_i(x) \in F(S).$$

In other words, $H(x)$ consists of all coalitions which can improve the outcome x. If the core is empty, then $H(x)$ is not empty for any $x \in X$. In this case we form the correspondence $G : X \Longrightarrow \Delta N$ by the formula

$$G(x) = \{\delta_S, S \in H(x)\}.$$

According to Proposition 3.6, there exists a bunch $Y \subset X$ and $g^* \in \mathrm{co}G(Y)$ such that the bunch Y is S^*-efficient (that is $P_{S^*}(Y)$ is empty, where $S^* = \mathrm{Argmax}(g^*)$).

Now if we add to the set of coalitions $H(Y) = \cup_{y \in Y} H(y)$ the singleton coalitions consisting of agents out of S^* we get the balanced set of coalitions

$$\mathcal{K} = H(Y) \cup (\bigcup_{j \notin S^*} \{j\}).$$

Now we consider the following family of subsets $A_i \subset X$:

$$A_i = \begin{cases} P_i(Y), & \text{if } i \in S^*, \\ X, & \text{if } i \notin S^*. \end{cases}$$

We assert that, for any coalition $S \in \mathcal{K}$, there holds $\cap_{i \in S} A_i \in F(S)$. This is obvious if $S = \{j\}$, where $j \notin S^*$, since in this case $A_j = X$. If $S \in H(y)$ for some $y \in Y$ then

$$\bigcap_{i \in S} A_i \supset \bigcap_{i \in S} P_i(Y) \supset \bigcap_{i \in S} P_i(y) \in F(S).$$

By monotonicity of F we have again $\cap_{i \in S} A_i \in F(S)$.

Due to the balancedness of F, we conclude that $\cap_{i \in N} A_i \neq \emptyset$. From the other hand, $\cap_{i \in N} A_i = \cap_{i \in S^*} P_i(Y) = P_{S^*}(Y)$ is empty due to S^*-efficiency of the bunch Y. □

As usually, a finite set X can be replaced by a compact set X at suitable topological conditions. Say that an effectivity function F is *compactly generated* if for any $Y \in F(S)$ there exists a compact subset $Y' \subset Y$ such that $Y' \in F(S)$.

Theorem 4.4 *Let X be a compact space, let F be a balanced and compactly generated effectivity function, and let the preferences P_i be acyclic and open in $X \times X$. Then the core is non-empty.*

Proof. First we prove the following lemma.

Lemma. *If $P_S(x) \in F(S)$ then $P_S(x') \in F(S)$ for all x' close to x.*

Proof of the lemma. Let $P_S(x) \in F(S)$. Since F is compactly generated we have $Y \in F(S)$ for some compact subset $Y \subset P_S(x)$. As P_i and, consequently, P_S are open, the set P_S contains the set $U \times Y$, where U is a neighborhood of the point $x \in X$. But the latter means that $Y \subset P_S(x')$ for any $x' \in U$. By monotonicity of F we conclude that $P_S(x') \in F(S)$.

We proceed to the proof of Theorem 4.4. In terms of the proof of Theorem 4.3, we obtain from the lemma that the sets $H^{-1}(S)$ are open. If the core is empty then the sets $H^{-1}(S)$ cover X. Choosing a finite subcovering and a closed subcovering in it, we get a closed non-empty-valued subcorrespondence $H' \subset H$. We form then the following closed correspondence $G : X \Longrightarrow \Delta N$, $G(x) = \{\delta_S, S \in H'(x)\}$. Afterwards all is completed as in Theorem 4.3, except that we use Theorem 3.9 instead of Proposition 3.6. □

Remark 4.5 The balancedness of an effectivity function F provides non-emptiness of the core at any acyclic preferences P_i. If now preferences P_i are fixed, non-emptiness of the core can be established at weaker conditions of balancedness, which appeal to the given P_i. Namely, one can see from the proof of Theorem 4.3 that in the formulation of the balancedness condition it is sufficient to take the sets A_i of the form $P_i(Y)$ (where $Y \subset X$) or X.

Taking into account Remark 4.5 we obtain, as a corollary of Theorem 4.4, the classical Scarf theorem about the core of balanced NTU-games (Scarf, 1967; Ekeland, 1979).

Acknowledgement This chapter is a translation from "Optimizatzija" 41(58), 36–49, Novosibirsk, by V.I. Danilov and A.I. Sotskov. Sasha, the second author, died tragically in an auto-crash on December 31, 2004.

References

1. Danilov V.I., and A.I. Sotskov (1998): "Maximal elements in topological convex spaces," Working paper 98.28, Cahiers ECO & MATHS, CERMSEM, Université de Paris-1-Pantheon-Sorbonne.

2. Ekeland, I. (1979): *Elements d'économie mathématique.* Paris, Hermann.

3. Kiruta A., A. Rubinov, and E. Yanovskaya (1980): *Optimal choice of allocations in complex social-economic problems.* Leningrad, Nauka (in Russian).

4. Polterovich, V.M. (1984): "Equilibrated states in the problems of vector optimization," *Automatika i Telemechanika*, 5, 89–96 (in Russian). See also his paper: "Equilibrated states and optimal allocations of resources under rigid prices," *Journal of Mathematical Economics,* 19 (1990), 255–268.

5. Scarf, H. (1967): "The core of an N-person game," *Econometrica,* 35, 50–69.

List of editors and authors

Vladimir I. Danilov, Central Institute of Economics and Mathematics, Russian Academy of Sciences, Nakhimovsky pr. 47, 117418 Moscow, Russia, vdanilov43@mail.ru

Gennadij N. Diubin, St. Petersburg Institute for Economics and Mathematics, Russian Academy of Sciences, Tchaikovsky st. 1, 191187 St. Petersburg, Russia, diubin@emi.nw.ru

Theo S.H. Driessen, University Twente, Department of Applied Mathematics, P.O. Box 217, 7500 AE Enschede, t.s.h.driessen@ewi.utwente.nl

Nikolai S. Kukushkin, Russian Academy of Sciences, Dorodnicyn Computing Center, 40, Vavilova, 119991 Moscow, Russia, ququns@pochta.ru

Gerard van der Laan, Vrije Universiteit, Department of Econometrics, De Boelelaan 1105, 1081 HV Amsterdam, glaan@feweb.vu.nl

Victor Lapitsky, St. Petersburg Institute for Economics and Mathematics, Russian Academy of Sciences, Tchaikovsky st. 1, 191187 St. Petersburg, Russia, victor_lapitsky@yahoo.com

Sergei L. Pechersky, St. Petersburg Institute for Economics and Mathematics, Russian Academy of Sciences, Tchaikovsky st. 1, 191187 St. Petersburg, Russia, specher@eu.spb.ru

Vadim I. Shmyrëv, Sobolev Institute of Mathematics, Siberian Branch of the Russian Academy of Sciences, Prosp. Akademika Koptyuga 4, 630090 Novosibirsk, Russia, shvi@math.nsc.ru

Valeri A. Vasil'ev, Sobolev Institute of Mathematics, Siberian Branch of the Russian Academy of Sciences, Prosp. Akademika Koptyuga 4, 630090 Novosibirsk, Russia, vasilev@math.nsc.ru

Elena B. Yanovskaya, St. Petersburg Institute for Economics and Mathematics, Russian Academy of Sciences, Tchaikovsky st. 1, 191187 St. Petersburg, Russia, eyanov@iatp20.spb.org

Fyodor L. Zak, Central Institute of Economics and Mathematics, Russian Academy of Sciences, Nakhimovsky pr. 47, 117418 Moscow, Russia, zak@mccme.ru

THEORY AND DECISION LIBRARY

SERIES C: GAME THEORY, MATHEMATICAL PROGRAMMING AND OPERATIONS RESEARCH

Editor: H. Peters, *Maastricht University, The Netherlands*

1. B. R. Munier and M. F. Shakun (eds.): *Compromise, Negotiation and Group Decision.* 1988 ISBN 90-277-2625-6
2. R. Selten: *Models of Strategic Rationality*. 1988 ISBN 90-277-2663-9
3. T. Driessen: *Cooperative Games, Solutions and Applications.* 1988 ISBN 90-277-2729-5
4. P. P. Wakker: *Additive Representations of Preferences.* A New Foundation of Decision Analysis. 1989 ISBN 0-7923-0050-5
5. A. Rapoport: *Experimental Studies of Interactive Decisions.* 1990 ISBN 0-7923-0685-6
6. K. G. Ramamurthy: *Coherent Structures and Simple Games.* 1990 ISBN 0-7923-0869-7
7. T. E. S. Raghavan, T. S. Ferguson, T. Parthasarathy and O. J. Vrieze (eds.): *Stochastic Games and Related Topics.* In Honor of Professor L. S. Shapley. 1991 ISBN 0-7923-1016-0
8. J. Abdou and H. Keiding: *Effectivity Functions in Social Choice.* 1991 ISBN 0-7923-1147-7
9. H. J. M. Peters: *Axiomatic Bargaining Game Theory.* 1992 ISBN 0-7923-1873-0
10. D. Butnariu and E. P. Klement: *Triangular Norm-Based Measures and Games with Fuzzy Coalitions.* 1993 ISBN 0-7923-2369-6
11. R. P. Gilles and P. H. M. Ruys: *Imperfections and Behavior in Economic Organization.* 1994 ISBN 0-7923-9460-7
12. R. P. Gilles: *Economic Exchange and Social Organization.* The Edgeworthian Foundations of General Equilibrium Theory. 1996 ISBN 0-7923-4200-3
13. P .J.-J. Herings: *Static and Dynamic Aspects of General Disequilibrium Theory.* 1996 ISBN 0-7923-9813-0
14. F. van Dijk: *Social Ties and Economic Performance.* 1997 ISBN 0-7923-9836-X
15. W. Spanjers: *Hierarchically Structured Economies.* Models with Bilateral Exchange Institutions. 1997 ISBN 0-7923-4398-0
16. I. Curiel: *Cooperative Game Theory and Applications.* Cooperative Games Arising from Combinatorial Optimization Problems. 1997 ISBN 0-7923-4476-6
17. O. I. Larichev and H. M. Moshkovich: *Verbal Decision Analysis for Unstructured Problems.* 1997 ISBN 0-7923-4578-9
18. T. Parthasarathy, B. Dutta, J. A. M. Potters, T. E. S. Raghavan, D. Ray and A. Sen (eds.): *Game Theoretical Applications to Economics and Operations Research.* 1997 ISBN 0-7923-4712-9
19. A. M. A. Van Deemen: *Coalition Formation and Social Choice.* 1997 ISBN 0-7923-4750-1
20. M. O. L. Bacharach, L.-A. Gérard-Varet, P. Mongin and H. S. Shin (eds.): *Epistemic Logic and the Theory of Games and Decisions.* 1997 ISBN 0-7923-4804-4
21. Z. Yang (eds.): *Computing Equilibria and Fixed Points.* 1999 ISBN 0-7923-8395-8

22. G. Owen: *Discrete Mathematics and Game Theory.* 1999 ISBN 0-7923-8511-X
23. F. Patrone, I. Garcia-Jurado and S. Tijs (eds.): *Game Practice.* Contributions from Applied Game Theory. 1999 ISBN 0-7923-8661-2
24. J. Suijs: *Cooperative Decision-Making under Risk.* 1999 ISBN 0-7923-8660-4
25. J. Rosenmüller: *Game Theory: Stochastics, Information, Strategies and Cooperation.* 2000 ISBN 0-7923-8673-6
26. J. M. Bilbao: *Cooperative Games on Combinatorial Structures.* 2000 ISBN 0-7923-7782-6
27. M. Slikker and A. van den Nouweland: *Social and Economic Networks in Cooperative Game Theory.* 2000 ISBN 0-7923-7226-3
28. K. J. M. Huisman: *Technology Investment: A Game Theoretic Real Options Approach.* 2001 ISBN 0-7923-7487-8
29. A. Perea: *Rationality in Extensive Form Games.* 2001 ISBN 0-7923-7540-8
30. V. Buskens: *Social Networks and Trust.* 2002 ISBN 1-4020-7010-1
31. P. Borm and H. Peters (eds.): *Chapters in Game Theory.* In Honor of Stef Tijs. 2002 ISBN 1-4020-7063-2
32. H. Houba and W. Bolt: *Credible Threats in Negotiations.* A Game-theoretic Approach. 2002 ISBN 1-4020-7183-3
33. T. Hens and B. Pilgrim: *General Equilibrium Foundations of Finance: Structure of Incomplete Markets Models.* 2003 ISBN 1-4020-7337-2
34. B. Peleg and P. Sudhölter: *Introduction to the Theory of Cooperative Games.* 2003 ISBN 1-4020-7410-7
35. J. H. H. Thijssen: *Investment under Uncertainty, Coalition Spillovers and Market Evolution in a Game Theoretic Perspective.* 2004 ISBN 1-4020-7877-3
36. G. Gambarelli (ed.): *Essays on Cooperative Games.* In Honor of Guillermo Owen. 2004 ISBN 1-4020-2935-7
37. G. B. Asheim: *The Consistent Preferences Approach to Deductive Reasoning in Games.* 2006 ISBN 0-387-26235-0
38. U. Schmidt and S. Traub (eds.): *Advances in Public Economics: Utility, Choice and Welfare.* A Festschrift for Christian Seidl. 2005 ISBN 0-387-25705-5
39. T. S. H. Driessen, G. van der Laan, V. A. Vasil'ev and E. B. Yanovskaya (eds.): *Russian Contributions to Game Theory and Equilibrium Theory.* 2006 ISBN 3-540-31405-9